Always On

how the iPhone
unlocked the
anything-
anytime-
anywhere
future—
and locked us in

Brian X. Chen

Da Capo Press
A Member of the Perseus Books Group

For Rachael

Editorial production by Marrathon Production Services, www.marrathon.net
Book design by Jane Raese
Set in 12-point Bulmer

Library of Congress Cataloging-in-Publication Data is available for this book.
First Da Capo Press edition 2011
First Da Capo Press paperback edition 2012
ISBN 978-0-306-81960-5 (hardcover)
ISBN 978-0-306-81982-7 (hardcover e-book)
ISBN 978-0-306-82076-2 (paperback)
ISBN 978-0-306-82210-0 (paperback e-book)

Published by Da Capo Press
A Member of the Perseus Books Group
www.dacapopress.com

Da Capo Press books are available at special discounts for bulk purchases in the U.S. by corporations, institutions, and other organizations. For more information, please contact the Special Markets Department at the Perseus Books Group, 2300 Chestnut Street, Suite 200, Philadelphia, PA 19103, or call (800) 810-4145, ext. 5000, or e-mail special.markets@perseusbooks.com.

10 9 8 7 6 5 4 3 2 1

contents

prologue

It's Friday evening, and as usual I'm meeting some friends, John and Rana, for dinner at 6. In a text message John suggests Alborz, a Persian place I've never been to, located in Pac Heights, San Francisco. The clock strikes 5:45 as I step into the elevator and launch the app Taxi Magic. I tap the "Book ride" button, and the app says a cab is arriving, about a half a mile from *Wired* headquarters. It tells me the driver's name is "Raj S." and the estimated cost of the fare is $12.

While I wait on the sidewalk for Raj to arrive, I launch the Yelp app to look up the address of Alborz. "1245 Van Ness Ave (between Sutter and Hemlock)." It's now 5:48, and I look up as a cab pulls in front of me. *That really is pretty damn magical,* I think to myself.

"Hey, Raj," I say as I open the car door. He looks mildly disturbed. "Can you take me to Van Ness and Sutter please?"

"Sure thing, boss," he says, with just a hint of an Indian accent.

My phone buzzes. Rana texts, saying she'll be a few minutes late (typical Rana). "No worries," I type in response.

"So where do you work?" Raj asks.

"Oh, I write for *Wired,* the technology publication," I chirp.

"*Wired!*" he says. "I love *Wired*! So that means you get to play with a lot of cool gadgets, huh?"

"Sometimes. I mostly cover Apple, actually, so I write lots of stuff about the iPhone. It's lots of fun—documenting how this technology is impacting us on a big-picture level."

Raj's tone of voice changes. "Okay—so bear with me for a second—you're a good guy to talk to about this."

I hit the power button on my iPhone to put it to sleep. "Uh huh?"

"So I think it's incredible that the iPhone has changed everything, and technology has done some pretty amazing things for us. But consider this: It's making us stupider than ever before."

I laugh. "Go on."

"No, really! It's making us idiotic. We rely on all this technology to tell us where things are, what to eat. We don't really know how to do anything on our own anymore. We're becoming anti-social, self-absorbed. We have all these problems that we create for ourselves. Bad reception, expensive phone bills. None of it is real!" He continues, "Now consider this: the Amish are the smartest people on the planet."

I laugh again, harder. "What?"

Raj goes on to explain that he's currently working to become an assistant professor in social science at San Francisco State, and for years he's studied the Amish. At the age of sixteen, Amish teenagers, he says, are given the freedom to leave their community to experience our "modern world" of sex, drugs, alcohol, and high technology. This experience is part of a tradition called "rumspringa," in which Amish teens can decide whether they wish to be baptized into the Amish church or to abandon the Amish life for our society.

"Ninety-five percent of them go back to their Amish life," Raj says. "You know why?"

"Well, I think it's natural to gravitate toward what you're used to," I retort.

"Sure, but what they're used to is a more wholesome lifestyle than what we have," he says. "These people have real human skills and real knowledge because they rely so little on technology. And they have real connections, real love, and even real problems because they're not communicating through all these digital barricades."

The light turns green, and we reach Van Ness and Sutter. On my iPhone, I punch in a $3 tip for Raj and press the "Pay" button, and I hear my receipt printing from the cab's machine. "Well, hey, I can't say I agree with you for the most part—and I'd argue with you if we had more time—but what you say about what we lose is pretty intriguing. I just need to think about it some more."

"The Amish," Raj repeats. "Really, look into it."

I thank Raj as I shut the car door behind me. It's six o'clock sharp, and I greet John inside Alborz. While we wait for Rana, I order a Cabernet and reiterate the conversation I had with Raj.

"That's absurd bullshit," he says. John, a forty-one-year-old iPhone software developer, has never been gentle with his words. "There are plenty of Amish who use cell phones, so they're hypocrites."

I chuckle. "Clearly he's generalizing, and if the cab ride were thirty minutes longer, the conversation would've been less silly. But what we lose—that other side of the coin—is certainly worth pondering on more, isn't it?"

Indeed, for months following that night I spent hours and hours conversing with friends and technologists about what we gain and what we lose in the iPhone future. As a technology news reporter for *Wired*'s website, every day I write a story about how the iPhone and the technologies it inspires are changing our world. But why stop there? What the iPhone and always-on gadgets can do today is fairly obvious; the far more fascinating question is, going forward, what does it all *mean*? How will this phenomenon change society and business? What will our world look like in a few years? And perhaps even more importantly, how is this revolution shaping each of us individually?

I realized the pros are about as fascinating as the cons are disturbing. The iPhone introduced the App Store, an experience in which you can instantly download and use new apps that add to the device's capabilities. With the tap of a download button, your iPhone can become a flute, a medical device, a high-definition radio, a guitar tuner, a police radio scanner, and 450,000 other "things." With the iPhone and the App Store, Apple unlocked what I call the anything-anytime-anywhere future, which has far-reaching implications for everything. If we have accessible data everywhere, then the way we learn in classrooms, treat medicine, fight crime, report the news, and do business are all going to have to transform.

For individuals, the iPhone is turning humans into always-on, all-knowing beings. Even without medical training, a person with an iPhone can use a first aid app to learn to treat a victim's injuries in an urgent situation. (In fact, a near-death earthquake victim in Haiti used a medical iPhone app to treat his wounds and, ultimately, survive.[1]) With the same device he can use a real-time traffic monitoring app to find the quickest route to a destination. Data has become so intimately woven into our lives that it's enhancing the way we engage with physical reality. Thus, the physical and digital worlds are coalescing to turn us into the super-connected beings we've always dreamed of being—and it took just one "phone" to push the industry in this direction.

Further, in the world of business, the benefits for consumers are fairly obvious. The iPhone changed our standards for what we expect from technology, and as a result, businesses are being forced to give us more for our money. We don't want seven pieces of hardware to perform seven different tasks; we want a single gadget capable of doing anything-anytime-anywhere. Soon, manufacturers will no longer be able to sell single-function gadgets lacking an Internet connection because those gadgets will be obsolete. Consequently, a large number of companies and industries find themselves threatened because a downloadable app can easily replace nearly any dedicated, single-use product.

But as ideal as it may sound to have anything-anytime-anywhere, the fact that Apple—a company famously obsessed with control—is leading this revolution is particularly concerning. Apple not only controls the manufacturing of the iPhone hardware, but it also oversees everything that appears in its App Store. Apple approves, rejects, or retroactively pulls any apps it pleases. This is comparable to if Microsoft not only sold you Windows but also owned every computer and every store in which it was sold and controlled every developer that wished to sell software for the computer. This sets a troubling precedent of censorship, which can stifle innovation and foster conformity. As technology becomes more intimately woven into our lives, the implications of this single point of control over our digital experiences are threatening creative freedom.

On top of that we must also consider what we give up as individuals in exchange for the incredible perks of anything-anytime-anywhere. Inevitably, the more we immerse our personal lives into digital media, the more privacy we give up. Businesses making apps have more information about our personal lives than ever before. Also, the application of basic civil rights is not keeping up with the rapid pace of high technology: police officers, for example, have the legal right to snatch our phones and look through all our personal information with "reasonable suspicion."[2]

Furthermore, after repeatedly sending text messages and e-mails in between checking Facebook and hopping on phone calls, looking in the mirror to ask ourselves, "What is the 'i' in iPhone?" is worthwhile; that is, how am I changing as a result of being bombarded with all this data? (I actually found myself asking this question a lot while writing this book as I was holed up in my office in front of a computer for a year.) Are we really getting stupider, like Raj suggests? The answer turns out to be much more complicated than Raj thinks.

Make no mistake: all the aforementioned implications go far above and beyond the iPhone. Everybody is copying Apple's closed, vertical business model in hopes of replicating the iPhone's success. Every major smartphone maker has rolled out iPhone clones and app store alternatives of their own, and their fundamentals (i.e., vertical control) are mostly the same. Apple's influence is even seeping outside the smartphone market. TV makers are already selling web-connected televisions, including app stores, and Ford will soon ship cars with app stores too—all with the common goal of trapping consumers inside their product lines.[3] Thanks to the iPhone, the future of business is looking vertical. Our products will enable us to do more than they ever have before, as their capabilities will be expandable with the tap of a download button. There's a tradeoff to allowing powerful, vertical companies to have so much control: We give up some individuality, creative freedom and, inevitably, some privacy.

Clearly, because it's impacting every facet of our lives, the future of anything-anytime-anywhere is unavoidable, making this a terrifyingly beautiful and exciting time to live. In an era when printed letters seem hopelessly limited when pitted against billions of minds posting on the Internet, this book is merely my attempt to paint a realistic portrait of our future with the help of some of the most intelligent technology thinkers, innovators, and researchers I've interviewed throughout my career. Let's explore together what being always on means.

Chapter 1

the dream
of the
perfect thing

The iPhone had everyone fooled. Even Steve Jobs didn't know quite what he had when he introduced the keyboard-less gadget in 2007.

"Today, we're introducing three revolutionary products," Jobs said during his keynote speech at Macworld Expo. "An iPod, a phone, and an Internet communicator. An iPod, a phone—"

He paused a beat.

"Are you getting it? These are not three separate devices. This is one device. We're calling it iPhone."[1]

Just three things? Talk about a tricky statement. As of this writing, the iPhone is hundreds of thousands of things and counting, thanks to "apps," which add to the handset's endless list of capabilities. The iPhone is not simply a web browser, a phone, and an iPod, but it is also a ballistics calculator for snipers, a barcode scanner for market mavens, a guitar tuner for musicians, a photo editor for shutterbugs, and much, much more.

In many ways the iPhone is the first gadget to come close to fulfilling our dream of the perfect device—the one that does it all—like Dick Tracy's radio-communication wristwatch or James Bond's lock-picking, fingerprint-scanning cell phone. Such is the undeniable appeal of a device whose minimal hardware disappears and, in the form of an app, becomes anything its owner wants.

But before Apple got there, on day one the iPhone didn't seem all that threatening to rivals (so they said, at least). Although Apple crammed into the iPhone a rich web browser, a state-of-the-art Maps application, and a redone iPod media player, plenty of competitors' phones sported similar features. What's more, the iPhone flashed an exorbitant price tag of $500—too little, too much, said critics.

"That is the most expensive phone in the world, and it doesn't appeal to business customers because it doesn't have a keyboard, which makes it not a very good e-mail machine," said Microsoft CEO Steve Ballmer, in a January 2007 interview with CNBC. "I, I kinda look at that and I say, well, I like our strategy. I like it a lot."[2]

Perhaps Ballmer missed the bigger point when Jobs introduced the iPhone. Even more significant than the phone itself was the business move behind creating it. Jobs slyly negotiated an arrangement with AT&T to carry the iPhone without even showing the carrier the device. This was unprecedented: the original idea was that a mobile carrier and manufacturer would determine the features they wanted on a phone and then they would issue a list of strict instructions to operating system makers. By directing the design and experience of the iPhone, however, Jobs wrestled control away from the carriers and, effectively Apple rewrote the rules of the wireless game. As a result, Apple was able to tightly control the design of the iPhone's OS and hardware in order to deliver a mobile experience tailored for the customer to enjoy rather than the carrier.

In the aftermath of the iPhone, by the third quarter of 2009 Microsoft lost almost one-third of its Windows Mobile market share compared with the third quarter of the prior year.[3] Humbled by the numbers, Microsoft would later admit to the weaknesses of Windows Mobile. Meanwhile, Apple's iPhone platform saw a healthy rise, from 12.9 to 17.1 percent. And by the end of 2010—at which point Apple had shipped 73.5 million iPhones[4]—Microsoft scrambled to announce its response: Windows Phone 7, an operating system that ran on—you guessed it— multitouch phones, some without keyboards. As of this writing, even after Windows Phone 7's release in late 2010, Microsoft is still a small fry in the smartphone game. The software company has even formed a partnership with Nokia, another crumbling giant that didn't react on time to the iPhone revolution, to make phones together in hopes of catching up. Critics wonder if the alliance will do any good, because the first Windows-powered Nokia phones won't hit the market until 2012.[5]

Untainted by carriers, the iPhone's consumer-friendly touchscreen experience was only the first part of what made the iPhone a big hit. In July 2008 Apple released the second-generation iPhone with a $200 price tag and a bubbly blue icon: the App Store. The App Store was the killer app that catapulted Apple ahead of its competitors.

The App Store struck a couple of chords. It gave iPhone users access to a wealth of third-party apps that developers from all over the world have coded. On day one the App Store launched with 552 apps.[6] By 2011 the App Store accumulated over 400,000.[7] By offering apps that filled every need, Apple retroactively delivered one device that can potentially replace any piece of hardware you could ever want to buy.

The iPhone unlocked a reality in which we can potentially have anything we want, anytime and anywhere. And as a result, everything has changed—from how people interact socially to how students learn in classrooms, and from how we do our jobs to how companies make products.

You have to wonder, what made the iPhone and the App Store so special? How did the iPhone unlock anything-anytime-anywhere? Wasn't this the dream—delivering to us anything digital we could possibly want, whenever and wherever we needed it—that the Internet has promised us for years?

It turns out that the web just wasn't enough—or perhaps it was too much. It depends on how you look at it.

The Way of the Web

After many years the Internet has evolved into a massive hub promising to grant access to every type of data we could possibly need. It is not by any means failing to fulfill that promise; the problem is that the Internet is *overdelivering*, and browsers are just too dumb. We have too much data now. We have so much data that we don't know how to make it useful. As a result, the Internet, constantly expanding like our universe, is so vast that search engines and web code aren't enough to integrate data into our lives in a manner that fits the human need.

The browser has become a dumb interface, but it was not always this way. During the 1990s Internet browsers were at the center of a firestorm of innovation when companies were dueling for domination of the web market.[8] Led by Jim Clark and Marc Andreessen, Netscape (originally called Mosaic Communications) pulled ahead with then-brand new HTML capabilities such as animation, audio, and video in the form of "extensions," which developers were churning out at an incredible pace. By 1995 Netscape owned upward of 80 percent of the browser market share; chances were good that if you were surfing the Internet, you were using Netscape.[9] And, of course, such big numbers intrigued none other than Microsoft, who launched a browser of its own, Internet Explorer, to challenge Netscape. Although Microsoft was years behind, it quickly gained market share because of one major distinction: it made Internet Explorer free from the start.

Netscape and Microsoft spent most of 1995 and 1996 duking it out with their browsers, issuing update after update and beta after beta. By version four of each browser, Goliath finally caught up and stomped on David. Netscape died in 1998 and was reborn as Mozilla—a free, open-source platform.[10] That was good news to Netscape fans, but many technologists agree that Microsoft's victory dramatically decelerated the development of web standards. After all, Microsoft's motivation for competing with Netscape was not to revolutionize the web but rather to gain an edge so as not to be left in the dust—this strategy has historically proven to be part of Microsoft's DNA. Since Netscape 1.0, however, the web-browsing experience hasn't changed fundamentally.

In the years that followed, very little innovation occurred in the browser space. With newer versions of Internet Explorer,

Microsoft mostly focused on upgrading security features. Not until the debut of Firefox much later in 2003 did the browser see some movement: version 2 of Firefox offered extensions, or add-ons, which enabled any coder to create new utilities to enhance the browsing experience. But after that, browser innovation plateaued—even when Firefox 3 released in 2008 one major "innovation" that Mozilla touted was "the awesome bar," a feature that automatically finds a previously visited site when you begin typing it in the address bar. Browser innovation stagnated, and the evolution of the browser experience has been slow relative to the rampant progress of the computers that we use to surf the web.

Joe Hewitt, one of the original creators of Firefox and former Facebook employee, recounted, "5 or 6 years ago web technologies completely stalled. It didn't go anywhere for the better part of the last decade until 2008 or so. We're still suffering from that six-year period where nothing was happening." He continued, "In the meantime you have Apple pushing this app store model on a platform they were able to iterate on. Once a year they add a ton of new stuff to the platform."[11]

But before we get to mobile apps, let's not forget about search engines. Search was a very important tool that made the web exponentially more useful, and it has come a long way. The Internet used to be a domain for academics, technologists, and the military, so finding information was a minor challenge then. However, between 1993 and 1996 the Internet saw a massive growth spurt, expanding from 130 sites to over 600,000.[12] The difficulty of search accumulated into a problem of epic proportion—one that Alta Vista, Yahoo, and, later, Google would tackle. John Battelle, cofounder of *Wired* and author of the book *The Search: How Google and its Rivals Rewrote the Rules of Business and Transformed our Culture*, sums up the Internet conundrum best: "That vastness is causing another kind of Web blindness: a sense that we know there's stuff we might want to find, but have no idea how to find it. So we search in the hope it will somehow find us."[13]

Search evolved from Matthew Gray's robot software called the Wanderer, which crawled the web and created an index of every site it found, to Google's large-scale hypertextual web search engine. Google eventually won the search game against AltaVista and Excite with its ranking algorithm, which started out as BackRub, a computer science project led by Larry Page and Sergey Brin at Stanford University. The duo found fascinating not only the practice of linking, but also the act of linking back. Page and Brin discovered that by studying how many sites link back to a particular site, we can assign a rank to how important that site may be—similar to how academic papers gain credibility when they're cited by other academic papers.[14] And beyond that, BackRub would also take into account the rank of the websites linking to a website to determine its rank. So rather than just spit out search results based merely on word strings, Google would sort results in accordance to their importance based on some clever math.

Those were just the early pieces of the Google algorithm that made it so much more sophisticated than its competitors. And as powerful and useful as search has become, the problem of search still remains only partially addressed. Udi Manber, vice president of engineering at Google, believes the search problem is only 5 percent solved, and that's just the nature of the beast: search will always be a question as much as it is an answer.[15]

Moving on to the mobile web, a new problem arises: search was a tool that was built for computers we used at home or in the office, and that's primarily where we use it most often. "Search traffic increases in the morning and peaks again in the evening, as we all fire up our home computers and look for movie tickets, homework help, or a local plumber to fix the dripping sink," writes Battelle.[16]

What about when we're walking down the street, driving cars, or shopping at brick-and-mortar stores? Suddenly, the equation changes. When we're outside or on the go, the slower connection and smaller screen of a smartphone inevitably cripples the act of search. Also, search is still a multiple-step process: type in a query, perform the search, choose the page, and load the content. For a mobile experience, search just isn't ideal. Although it magnificently improved the Internet-browsing experience on computers, for the always-on lifestyle unlocked by a smartphone, there were plenty of ways search could be better. Search wasn't smart enough for the mobile experience, and this was one major area that Apple improved with native apps.

What's more, web browsers got even dumber when they moved to smartphones. They were unable to support many of the rich web features to which we've grown accustomed for several years on our computers. Given those limitations, web developers dumbed down the mobile versions of their websites—if they even bothered to make them at all. In response, Apple was the first to introduce a rich HTML mobile browser with the first iPhone, which was a significant move because it inspired many web developers to repurpose their websites just for the iPhone (and eventually its copycats)—and the mobile web then quickly matured.

"It was Mobile Safari and the iPhone that was what made the mobile web infinitely more useful," said my colleague Michael Calore, editor of Wired.com's Webmonkey, who has studied and reported on the web space for several years. "Before that, it was largely utilitarian, or a novelty even."[17]

In 2007 Jobs announced that developers could code web apps for the iPhone, and this seemed like good news. But a rich web browser and an operating system that ran web apps still weren't enough: months passed, and barely anybody coded web apps. Why? There wasn't good money to be made—there was no cohesive business platform on the web or through a browser that made web apps a lucrative opportunity. (Hobbyist programmers already learned that the hard way when they took a stab at the frivolous shareware market.)

To make the iPhone platform special—and competitive— Apple needed to recruit developers to make third-party apps. After all, that's part of how Microsoft locked down its dominance of the desktop OS space: the vast majority of software developers made games and apps for Windows, not the Mac. Apple needed to rally developers for the iPhone. And in order to do that, Apple had to offer an incentive.

Jobs had just the thing.

A Business Model That Just Worked

"So you're a developer and you've just spent two weeks, and maybe a little longer writing this amazing app, and what is your dream?" an energetic Jobs said to a packed Apple headquarters conference room of software developers and journalists in March 2008. "Your dream is to get it in front of every iPhone user. That's not possible today. Most developers don't have those kinds of resources. Even the big developers would have a hard time getting their app in front of every iPhone user. We're going to solve that problem for every developer, big to small."[18]

Jobs was swinging his arms more than usual—reminiscent of when Willy Wonka, played by Gene Wilder, was welcoming people into his chocolate factory. I could tell he was excited.

"The way we're going to do it is what we call the App Store," he said. "This is an application we've written to deliver apps to the iPhone. And we're going to put it on every single iPhone with the next release of the software. And so our developers are going to be able to reach every iPhone user through the App Store. This is the way we're going to distribute apps to the iPhone."

And it worked. Developers did make money—some more than others. A few even struck gold: independent coder Steve Demeter in September 2008 said he earned $250,000 in just two months with his puzzle game Trism, which he sold for $5 a copy. Perhaps the most enticing part of Demeter's story was that Trism epitomized "indie-ness": he made the game almost entirely by himself, with a little help from a contracted designer to whom he paid $500, and he publicized the game by himself with Twitter and other social networking tools.

Being a journalist who isn't used to seeing big paychecks, I gawked when I looked over Demeter's bank statement, which he presented to me as proof of his success: at the end of the month was a single transaction of $90,000 from none other than Apple. The best part, Demeter bragged, was that he wasn't even trying to make money.

"I really didn't think about the money," Demeter told me. "I got an e-mail from a lady who's like, a fifty-year-old woman, who says, 'I do not play games, but I love Trism.' That's why I did it."[19]

Demeter became the poster child for the app story. Apple piggybacked off Trism's success, highlighting its hot sales in a press release and even a film documentary that was shown during a press event.

Six months later, in January 2009, an even greater success story surpassed Demeter when independent coder Ethan Nicholas earned $600,000 in a single month with his tank-artillery game iShoot. Nicholas's story was even more romantic. After getting off his shift as an engineer at Sun Microsystems, he worked on iShoot eight hours a day, cradling his one-year-old son in one hand and coding with the other. He didn't have the money to buy books to learn how to write an iPhone app, so he taught himself by reading websites.

When iShoot launched in October, business was slow for a while. And then Nicholas found some spare time to code a free version of the app, iShoot Lite, which he released in January. Here's how that helped: Inside iShoot Lite he advertised the $3, full version of iShoot. Users downloaded the free version 2.4 million times, which led 320,000 satisfied iShoot Lite players to pay for iShoot. The game soared to the No. 1 spot in the App Store's list of bestsellers, and it stayed there for twenty-six days. The day iShoot hit number one, Nicholas quit his job.

"I'm not going to be a millionaire in the next month, but I'd be shocked if it didn't happen at the end of the year," Nicholas told me when iShoot was still number one. "If it weren't for taxes I would be a millionaire right now."[20]

As hopeful as Demeter and Nicholas were when they both spoke to me, months later their success turned out to be mostly luck after all. In an October 2009 interview with *Newsweek*, Demeter said that after Trism sales slowed down, he only really made big money after investing his App Store earnings in the stock market. Similarly, Nicholas hasn't come out with a big hit like iShoot since, and he told *Newsweek* that he was terrified of being a one-hit wonder.

The press, including yours truly, was quick to label the mobile app opportunity a digital gold rush. Some less successful developers felt this was a sensationalist generalization, though it seemed a fair analogy. During the California Gold Rush in the mid-nineteenth century, only the lucky few struck it rich, but they attracted hundreds of thousands of eager Forty-niners from across the continent. Likewise, only a small number of iPhone developers struck digital gold, whereas others panned nuggets or didn't gain much at all. Lured by the dreams of riches, hundreds of thousands of programmers enthusiastically signed up to produce iPhone apps.[21]

With the dot-com boom gone bust, the App Store birthed a new digital frontier. And with giants dominating Silicon Valley, start-ups and independent programmers soon realized they could fit in between the cracks by coding mobile apps. They delivered nearly 400,000 in less than three years.[22]

The Genius of the Blank Slate

The iPhone took Apple's core belief—that software is the key ingredient to hardware's success—and expanded it. Apple fashioned the iPhone as a blank slate with only one button and a touchscreen, whose special powers lay beneath the operating system. The customizable, intuitive interface combined with the wealth of third-party apps available for the device made the iPhone the first phone of choice for any type of user: consumers, professionals, teachers, students, doctors, and even the US Army, which is experimenting with iPhone apps in the field. What's more, Apple didn't exclude younger, less wealthy customers, either: the company released the iPod Touch, best described as a phoneless iPhone, which required no monthly phone bill. Apple's expansion of the market amounted to over 120 million iPhone and iPod Touch customers by 2010.[23]

iPhone apps addressed a number of issues that the web and search posed in mobile applications. The fact that iPhone apps were made only to run on the iPhone meant that they were optimized to take advantage of the iPhone's own processor, which, among other features, included GPS. As a result, apps were faster at retrieving data and they were smarter at pulling information in a manner specific to a particular app. (The Yelp app, for example, searches only the Yelp database for nearby restaurants and entertainment venues based on the geographical location of your iPhone.) Furthermore, unlike computer apps, iPhone apps maintain themselves by automatically running and installing software updates with the tap of a button, whereas PCs have to navigate to websites, download files and keep software updated manually. Thus, everything about the iPhone and the App Store was designed to fit an always-on, mobile lifestyle.

So Apple did not invent something brand new out of necessity and thereby lift off a revolution. After all, so much of what apps can do has been doable for years. Instead, Apple's modus operandi has traditionally been to study an existing technology and determine how it can be done better.

"Apple doesn't necessarily invent things so much as reinvent them," said Matt Drance, a former software engineer at Apple who helped evangelize the iPhone platform during its infancy. "I think it's the same thing with mobile. Certainly Apple did not come up with the first phone. They sat and watched for a long time, and they finally took what they observed and what they learned when deciding to solve the problem."[24]

That's also what Apple did with the Apple II in 1977. Instead of delivering a PC that required the user to assemble circuit boards and soldering irons, which the rest of the PC market was doing,*[25] Apple packaged all the ingredients of a functioning computer into a convenient plastic case. An Apple II was ready to use the moment after it was plugged in; in so doing, average people, not just hardcore geeks, could now use the personal computer. The story is practically the same for the iPhone: Apple delivered a phone designed for customers to enjoy along with an App Store where they can discover tools tailored to their needs in a friction-free way.

Thus, the iPhone's App Store was a blockbuster hit. There are plenty of websites that replicate what many iPhone apps can do, but customers found that iPhone apps could do them better, and they were crazy about downloading and even paying for them. The App Store hit a billion downloads in April 2009, and by January 2011 it had surpassed ten billion.[26]

Numerous observers argue that 400,000 apps is insignificant because a lot of those apps are garbage, ranging from fart apps to really lame games.[27] And the apps that make the iPhone stand out, the ones with stellar quality, are few and far between. (The popular Twitter app Tweetie is often hailed as a prime example for a piece of software coded with beautiful design and rich features.) Quality apps were extremely important for the iPhone's success, but quantity was just as vital, maybe more so.

The more apps the App Store accumulates, the higher the chance the App Store has to appeal to each of the millions of iPhone and iPod Touch owners in the world. We can ignore a

---

*The exception was the Commodore PET, which was the first to sell an all-in-one computer the same year that the Apple II debuted.

horde of lousy iPhone apps, and there are many low-profile apps that average consumers wouldn't pay attention to or care to use. But there's also a plethora of niche apps catering to specific professions, hobbies, and interests—those apps that slip past the average consumer's radar. Some examples include iChart EMR, an app for doctors to view and store patients' medical charts; Rev, an app for mechanics to perform car-engine diagnostics; and Nerdulator, an app for military snipers to calculate ballistics. These kinds of niche apps are what make the iPhone special. Consequently, with the help of quantity as much as quality in the App Store, Apple delivered a device that comes close to fulfilling our dreams of the perfect device.

However, Steve Jobs didn't just look at the problems of the web and magically pull a solution out of a hat; rather, Apple spent years tinkering with different software and distribution models on the Mac before it nailed the sweet spot. The App Store wouldn't be anywhere near as successful as it is today without iTunes.

The Birth of iTunes

Between the mid-1980s and late 1990s the media were undergoing a massive conversion from analog to digital. The music industry hated it.

Much to the chagrin of the Recording Industry Association of America, Internet users quickly caught on to digital music as a free alternative to paying for albums, thanks in large part to the WinAmp MP3 player and Napster. Facing declining album sales, record labels filed lawsuit after lawsuit against online services Napster and MP3.com for hosting digital music, as well as Diamond Multimedia, a Korean company that released an MP3 player called the Rio. Clearly, for the recording industry, change wasn't easy.

In stepped Steve Jobs. The Apple CEO harbored a vision in 2002 of an online music store hosted by Apple that would be easy to use, complete in selection, and reliable in performance. These factors, Jobs thought, would be enough to convince customers to pay for something they could otherwise obtain for free illegally. The store, then, would enable record labels to compete with pirates rather than pursue a futile attempt to destroy them.

However, in order for online music to work, Jobs believed his store would have to allow customers to purchase music in a completely different way: à la carte. Convincing labels was not easy. "When we first approached the labels, the online music business was a disaster," Jobs told Steven Levy, author of *The Perfect Thing*. "Nobody had ever sold a song for 99 cents. Nobody really ever sold a song. And we walked in, and we said, 'We want to sell songs à la carte. We want to sell albums, too, but we want to sell songs individually.' They thought that would be the death of the album."[28]

Jobs started his talks with the big players first: Warner Music and Universal. Apple flew the firms' teams up to Cupertino, California. In a boardroom at One Infinite Loop, Jobs proceeded to present his plan. First, he reeled in the labels with one crucial proposal: Apple would sell songs through iTunes, music-player software that was then available only for Macs. After all, how could Apple, whose Mac operating system held only single-digit market share, ruin the record business if the iTunes Store took off?

After a series of long and painful negotiations, the two labels ultimately agreed they would play, but only after Apple agreed to some restrictions: iTunes-purchased songs would be limited to being playable on three "authorized" computers, and a playlist could only be burned on a CD seven times.

Labels BMG and EMI soon followed, and later Sony hopped on board. Apple opened the iTunes Music Store on April 28, 2003, with 200,000 songs. (Simultaneously, Apple released its third-generation iPod.) In the first week iTunes Store customers bought more than a million songs. Six months later Apple convinced the labels to allow iTunes to be shared with Windows users.

The iTunes Store was a prequel to the anything-anytime-anywhere experience that the iPhone would later unlock. The most important feature it used to compete with pirates was the ability to download a song from a huge digital catalog with the simple click of a button—an experience that the iPhone expanded vastly. The rest of the story of how the iPhone unlocked the anything-anytime-anywhere revolution is best illustrated by some of the most successful independent programmers involved in the app scene.

Chapter 2

a new
frontier

When he was thirteen years old, Phillip Ryu and his father Seungoh scooted desks side by side to work on their first project together: a Mac version of Pong. They finished the game over a weekend and then slapped it on their website for $5 a copy. It was just a playful experiment, but to the Ryus' surprise, their Pong game was a hit among Mac users and some popular blogs.

It was 2001, and Apple had just released its brand-new operating system, Mac OS X. There was barely any software out for it, so in retrospect the fact that the Ryus' Pong game became popular, crude as it was, wasn't shocking. "We were staring at a new frontier in the face and didn't recognize it then," Phill told me. "When you start out with no competition, the sky's the limit."[1]

The Korean father-and-son company continued producing Mac apps. Their next creation would be an app called iStorm, which enabled multiple people to collaborate on a document in real time over an Internet connection. The app was the first of its kind for the Mac—and, again, a modest hit. With their small software projects, the Ryus would earn $5,000 to $10,000 a year to spend on family vacations.

However, as the Ryus' success grew, their business relationship deteriorated. Seungoh, a research physicist at the time, was doing all the coding for the projects; Phill only knew HTML programming, so he directed interface design and managed the website to market the apps. Their partnership came to an end with a dispute over button colors: Phill wanted the buttons in iStorm to conform to the "Aqua" (blue and white) color scheme of Mac OS X, whereas Seungoh favored a metallic gray aesthetic. "I would nag at him to do something with user interface, and he wouldn't implement it. And we kept doing that until he eventually fired me," Phill said.

So when Phill was fifteen, he and his father split as business partners, but that didn't end the story for the young dreamer. About two years later Apple released an upgraded version of Mac OS X, dubbed Tiger, with a new feature called the Dashboard. The Dashboard would be essentially a screen that always ran in the background "widgets," or lightweight apps such as a calculator or weather checker that could be accessed with a keystroke (F12). Programmers could make their own Dashboard widgets and put them up for sale online, and Apple would handpick one widget a day as a featured item on Apple.com. With this, Phill saw another frontier of opportunities.

Collaborating with web programmers and designers he met online over years in the Mac community, Phill got to work on some widgets. His first popular seller was VoiceNotes, a widget for recording voice memos, priced at $6. And after a few more hits (and some misses), by the time Phill began his first year of college at Dartmouth, he and his team had earned about $30,000 just with widgets.

Not bad for a group of teens, who would've taken ages to earn the same amount flipping burgers or brewing espresso for minimum wage. But Phill still saw a big problem. Only about eight thousand people—a small fraction of the overall Mac population—were buying his widgets. Therefore, even with the occasional help from Apple's promotion, websites were still a grossly inefficient business model for getting people to buy software. Phill believed that if there was a more centralized, streamlined business model, programmers could produce better, more niche applications and then sell them at lower prices, thereby reaching more customers. With the website-and-shareware model, however, a programmer could, for example, spend a fortune making a special widget for auto mechanics to diagnose vehicles, and barely anyone would buy it. There weren't enough customers to make up for development costs—so why bother doing anything cutting-edge? To make a long story short, digital distribution was limited and chaotic, and this stifled innovation. "Within the Mac shareware space there was always a healthy indie development community—a lot of hobbyists mostly having fun making cool little tools that they want for themselves," Phill said. "It was always a cottage industry. What I became obsessed with was how many Mac-using friends are actually buying these apps."

For the years following, Phill focused on the flawed, decentralized business model of websites and shareware. He wondered how someone could make software distribution more efficient. In an effort to work on a solution for the problem, he joined forces with John Casasanta and Scott Meinzer, software developers he had worked with in the Dashboard scene. Together they launched a bizarre project called MacHeist.

A white Subaru WRX chases a silver Mercedes out of a parking struc-ture and down a country road. They pull over outside a brick build-ing, and the target steps out of his car. He's wearing charcoal New Balance sneakers, dark blue jeans and a Jobs-ian black turtleneck. He pulls his weapon out of his pocket—an iPhone.

That sounds like a scene from a geeky James Bond parody, but it was actually the video conclusion to the third MacHeist software sales stunt. The annual MacHeist event consists of a se-ries of online missions, with clues hidden inside videos and miniature games to help participants solve puzzles, and the com-pletion of each stage unlocks access to a free Mac application. This gimmicky scheme is designed to promote the sale of a Mac software bundle containing about a dozen applications that in-dependent programmers have made. The bundle is priced at about $40—with each app in the bundle normally costing that much alone, thus making it an irresistible sale for Mac users.

"We were hacking the consumer decision process, whereas instead of buying these $40 to $50 apps one at a time, you can have them all in a bundle for $40," Phill explained. "We tried to make it a nondecision as much as possible—like you're an idiot if you don't pick this up."

MacHeist's tricky puzzles and low price tag stirred some seri-ous buzz in the Mac community. Overall, MacHeist 3 sold ninety thousand bundles, totaling more than $3 million. The MacHeist organizers split the earnings between themselves, participating programmers, and charities.

Despite its phenomenal success, however, MacHeist under-scored a major challenge that independent developers faced: the struggle to get attention for their apps in an increasingly cloudy ecosystem of Mac software. MacHeist's elaborate marketing stunt, which costs the team about $400,000 in marketing each year, illustrates how much of an effort is required just to get a sizable chunk of the Mac population to download software.[2]

To Phill, the problem of inefficient software distribution still remained unsolved. He teamed up with friend and programmer Austin Sarner to work on a better solution, and they came up with a piece of software called Project Serenity. The software would be a free application manager with which Mac users could download and run new apps from a centralized store. In other words, Ryu and Sarner were aiming to make a Mac app store. However, only two to three weeks into their progress on Project Serenity, Steve Jobs announced the upcoming iPhone App Store. Knowing that a Mac app store was on the horizon,* the pair shelved Project Serenity.

"Every problem and inefficiency of the previous system was no longer an issue in this new App Store model," Phill said. Consequently, rather than obsessing over fixing the distribution model of software, Phill now returned his attention to developing software—for the iPhone, of course. Phill paired with programmer Andrew Kaz, whom he met through Sarner, to code one of the first iPhone e-book readers, Classics, priced at $3 in the App Store. In its first year the app sold about 283,000 copies, earning roughly $850,000. After Apple took its 30 percent piece of the pie, Ryu and Kaz raked in a profit of over $600,000 for the app.

Today, the App Store, now nearly three years old, is only in its infancy. Phill foresees the app phenomenon becoming the sequel to the dot-com boom—a frontier with a bright future in light of the rise of smartphones and tablets.

---

*Apple announced the Mac app store in late 2010.

"I grew up hearing stories from my dad about the early days of the PC revolution, back when things were still moving at a lightning pace, and today's software giants were starting in garages. Everything was possibilities," said Phill, now twenty-two and living in San Francisco. "This is all happening again right now in the mobile apps industry, and we are writing this second chapter. This is it."

Boom

First he made an island. Then he added a sun, clouds, and a body of water. Finally, he sprinkled in some life: a seagull, a fish, a shark, and human-like creatures called Pygmies. Like a god, David Castelnuovo had created a miniature world in an iPhone game. Appropriately, he named it Pocket God. "I've always had an entrepreneur streak, and I wanted to have control over things I did, and this was basically the pinnacle," said Castelnuovo.[3]

The iPhone was a godsend to Castelnuovo (pardon the pun). For several years Castelnuovo had programmed games for major game companies, including Sega, but eventually he burned out. In 2001 he started his own company, Bolt Creative, a development studio that specialized in making Adobe Flash games and apps. However, it didn't make much money, and he didn't find management to be rewarding.

So when the App Store opened in 2008, Castelnuovo's dream came true. He had always wanted to self-publish a game. He quickly learned Objective-C and Cocoa, the programming languages used to make iPhone software, and he whipped up Pocket God in only a week. Castelnuovo intended the project to be practice for a more serious, "real game," but Pocket God became huge.

Pocket God initially gained popularity with barely any promotion or media coverage: pure luck and sheer customer satisfaction nurtured its organic growth. "The game was just really accessible with no rules, no guidelines, and easy to whip out at a bar and show it off on your phone," he said. "Apple tapped into the real casual gaming market." After only two months, his independent creation soared to the number-one spot in the iPhone App Store's list of best sellers, and within one year his game earned him over a million dollars. By mid-2010 Pocket God had surpassed $3 million in sales.[4]

Croatian brothers Igor and Marko Pusenjak have an even better success story. They saw the App Store as a playground for experimentation. Igor, who has an MFA in multimedia design, was living in Chelsea, New York, making websites; Marko was still juggling engineering gigs in Croatia, earning a meager $1,000 per month and wanting to make some extra dough. They started out with a novelty app simulating bubble wrap (iBubbleWrap), and after they raked in a few thousand bucks with that, they took the App Store more seriously. They experimented with making software for kids—to start, a silly game with a rabbit catching carrots—and later they sketched together a charmingly simple creation called Doodle Jump.

Doodle Jump starred a small alien-like creature that hopped from platform to platform. All an iPhone user had to do was tilt his or her hand to move the creature left and right as it jumped higher and higher, being careful not to let the creature fall. The entire game looked like it was doodled on graphing paper—an aesthetic that screams nostalgia for adults and, well, cuteness for kids. Doodle Jump was also a viral success, leaping higher and higher into the top-selling charts and, in three years, earned $5 million in revenue for the two brothers. "The level of entry was much lower [than other game platforms]," Igor told me. "We weren't in it to speculate and make boatloads of cash. We were a two-guy shop with a good idea of what it takes to program a great game."

Australian programmer Rob Murray's story was a bit of a wild ride. In 1999 he left a game development studio to start his own mobile game company, Firemint. When the App Store hit the scene in the summer of 2008, Murray and his team went to work on a driving game for the iPhone, Real Racing 3D. The team spent over $1 million on the project. Then, when Christmas break rolled around, Murray decided to experiment with a casual project for fun. "I loved the feeling of drawing on the iPhone screen, so that was something I really wanted to play with," he told me. "I needed a setting that fit, and I researched professions that seemed exciting and a little glamorous in some way. I was really intrigued by air traffic control. I kept reading about how it is an incredibly stressful job but how many of those who do it absolutely love it."[5]

Within one week Murray created a prototype for a cartoonish game in which iPhone users draw lines with their fingers to land aircraft with the objective to land as many as possible while avoiding collisions. With some help from his wife, Alex Peters, who wrote the text and sourced the sound effects and music, along with art director Jesse West, who refined the art into a retro look, he finished the game by February. They called it Flight Control.

Much to Murray's surprise, Flight Control was a megahit, selling over two million copies at $1 a download in its first six months. Real Racing was also profitable, though Murray declines to disclose sales numbers. "I imagine it's a bit like having one child who grows up to be a movie star and another who grows up to be an astronaut," Murray told me.

These stories illustrate that Apple's App Store gave birth to a new digital frontier by pioneering a business model in which software developers both big and small had an equal chance to make good money. What mattered most was the idea of an app and how its creators executed it. Beyond that, long-term success depended on marketing and other strategies, such as selling features within the apps. The App Store opened the floodgates for innovation from all sorts of brilliant minds who were creating games and other apps that appealed to people of all ages. Apple did what no other company had done before: it brought software into the mainstream, bringing together people who shared an intimate mobile experience. iPhoners everywhere were whipping out their iPhones and showing off apps and games to one another—since when did people do that? "People never talk about software or the new web startup unless you're actually in the industry, but it's become so pervasive now that Average Joes get the basic concept," Phill said. "Friends are recommending apps to each other, and that's just crazy. We're nowhere near as big as this can be."

The Competition

Apple and iPhone programmers weren't the only people to benefit from the app phenomenon. Following Apple, Google launched its Android mobile platform and opened an app store of its own in October 2008, offering the same 70–30 revenue split as the Apple App Store. The Android apps market grew to 150,000 apps after about two and a half years.[6] The Android app store saw smaller, but still notable, success stories than the iPhone's.

Twenty-six-year-old Stanford graduate Eddie Kim picked up a book about coding Android apps while traveling in Japan because he thought it might be a fun hobby, and six months later his casual creation would earn him more money than any of his full-time jobs. Kim, a former Volkswagen engineer and cofounder of San Francisco–based startup Picwing, earned $13,000 in just one month with his Android app Car Locator, which helps users find their parked cars. Kim's app, which he sells for $4 per download, took him only three weeks to code. "I thought about making an app for the iPhone, but my thoughts were, it's such a crowded space right now, and I thought Android would be a better opportunity to get involved in," Kim told me.

However, the lack of a simple, centralized market for third-party apps stifles Android developers. Google's phone strategy is the opposite of Apple's: the search giant hands out its operating system to any manufacturer willing to run it, and, thus, there are multiple phones running various versions of Android, each supporting different apps from the Android store. Some developers have shied away from the Android platform, turned off by "hardware fragmentation"—having to develop and support several versions of the same app for various kinds of phones from several manufacturers. By comparison, the iPhone offers a relatively clear-cut audience of one hundred million iPhone, iPad, and iPod Touch customers, with smaller differences in features between the various models.

Of course, Microsoft wasn't going to be left out of this game. In December 2008 the software giant scrapped its entire Windows Mobile project and started over—an event that employees allude to as "The Reset." [7] The company spent six weeks hatching a plan for a Windows phone do-over, and it set a deadline of one year to build and ship a brand-new mobile OS. The Reset was no simple task: it involved bringing in new managers, reorganizing the Windows phone-design department, and opening new test facilities dedicated to mobile hardware. The end result would be a consumer-oriented touchscreen phone, Windows Phone 7, which finally hit the market late 2010. Like the iPhone, Windows Phone 7 launched with an app store expanding the smartphone's capabilities with third-party software and games. Again, Microsoft offered the same 70 percent cut to developers for app sales.

Like it did before with Windows Mobile, Microsoft offered the Windows Phone 7 software to manufacturers willing to support the operating system. This time, however, Microsoft set some rules: each device running Windows Phone 7 must meet specific hardware criteria (three standard hardware buttons, for example), and before they can be shipped with Windows Phone 7, the phones have to pass a series of quality-assurance tests performed by Microsoft-made robots that check touchscreen calibration, sensor sensitivity, and more. This process ensures a consistent user experience across different phones, which in turn should enable developers to create apps more easily than they would on Android. Although Microsoft still has a lot of catching up to do, with this tightly controlled model, Windows Phone 7 has the potential to surpass the Android platform in the next few years in both app development and customer base.

"The success of the iPhone certainly had an impact on the industry and an impact on us," said Joe Belfiore, Microsoft's corporate vice president of Windows Phone. "And we said there were a lot of things we could do to deliver a solution that's different from the iPhone but have some of its benefits."[8]

Above and Beyond Phones

The app store model is reshaping the future of business—and it doesn't stop with phones. Any piece of hardware with a screen can be both enhanced and future-proofed—or continually

upgraded in the future with new apps and software updates—
with an app store. "We've got multiple connected screens,
whether it's the screen in your car, screen in your pocket, or
screen in your living room," said Michael Gartenberg, a technology strategist at Altimeter Group. "The Internet is connecting
these devices together and information is flowing from screen to
screen. Beyond that it's not just content but functionality. The
key is matching the application functionality to the screen and
function. . . . Whoever gets those applications and the services
into place first is going to control a lot of the market."[9]

For example, TV makers are beginning to ship television sets
with Internet connectivity and web channels offering additional
content—basically TV app stores. Toshiba's Cell TV, for example, features supercomputer processors, a video phone, and a
web channel to access both Internet-streamed content and, later,
apps—again, an anything-anytime-anywhere experience.

Furthermore, Ford is planning web-enabled cars equipped
with screens that will have access to an app store. Examples of
car apps include tour guides that work with your GPS, software
that promotes greener driving by giving you points for driving at
less erratic speeds, and, of course, web-streamed music from services such as Pandora. Ford says it will allow third-party programmers to create apps for the store. If more car makers follow
suit—and they probably will—the car industry will provide another potential frontier of opportunities for programmers. "With
the rise of smartphone platforms and the television becoming a
more intelligent portal to all kinds of information, it's been a lot
easier to gain distribution to a consumer base than it has been
previously," said Ross Rubin, a consumer technology analyst
with NPD. "As these platforms proliferate, it's going to buy the
developer exposure to many more customers."[10]

Console makers are already up-to-date. Every major console, including the PlayStation 3, Xbox 360 and Nintendo Wii, each feature online game stores—though they still primarily ship physical media, presumably to keep game prices high.

And let's not forget about tablets. Apple's iPad, launched in April 2010, opens a host of new opportunities for digital content. Most interesting to witness is how it will change the publishing industry (magazines and newspapers) and education (textbooks and tools for learning and research). Major publishers, including *Time*, *New York Times*, and my parent company, Condé Nast, are actively experimenting with iPad apps, and several universities, including Seton Hill University, George Fox University, and Abilene Christian University, have already started pilot programs using the iPad as an all-in-one textbook. "Those big, heavy textbooks that kids go around with in their backpacks are going to be a thing of the past," said Mary Ann Gawelek, vice president of academic affairs at Seton Hill, which gave iPads to its twenty-one hundred students and three hundred faculty members in the fall of 2010. "We think it's leading to something that's going to provide a better learning environment for all of our students. We're hoping that faculty will be able to use more of a variety of textbooks because textbooks will be a little bit less expensive."[11]

We're living in the age of the all-in-one connected device—the era of hardware consolidation. The iPhone unlocked this era, in which hardware manufacturers are competing with platforms that promise to offer potentially anything, anytime and anywhere, all with the click of a button.

As a result, the app store craze is coming full circle—as in, back to the PC. In October 2010 Apple's Jobs said he was taking lessons learned from the iPhone and iPad and applying them to the new MacBook Air, Apple's ultra-thin notebook, which weighs less than two pounds and sports long battery life. In addition he announced the next Mac operating system, Mac OS X Lion, which would ship with—you guessed it—a Mac app store. Just like the iPhone and iPad, Macs running Lion would gain access to an app store for third-party software. Apple again would split revenues with programmers 70-30. "It's going to be the best place to discover apps," Jobs said.[12]

Though this new Mac app store is simply an expansion of an already-done idea for Apple, it signified a radical shift for the way PC software is sold, said Forrester Research analyst Sarah Rotman Epps. "Software doesn't come in boxes anymore," she said. This new way of getting software could drive down its prices while also making it more accessible to the masses. Indeed, some programmers think a similar anything-anytime-anywhere revolution is on the horizon for the Mac. Just as the App Store did with the iPhone, we can expect a wave of new programmers opting to make apps for the Mac, and as a result, customers will get thousands of Mac apps, thereby enabling their computers to do things they never even thought about. "I think it can breathe some new life into Mac software," said John Casasanta, partner of the MacHeist software bundle.[13]

Looking further ahead, you can bet that competing PC manufacturers will be adopting app stores as well. Leaked company documents suggest that Microsoft is planning to introduce an app store in Windows 8.[14] Google has likewise announced plans for Chrome OS, a browser-based operating system running on small notebooks, and this will be launching in 2011 with a web-based app store.

By giving software developers a centralized business platform from which they can cultivate entire companies and careers, Apple has empowered content creators more than any other electronics company and, as a result, gained more capabilities than ever before. To some extent every programmer is a digital god of his or her own creation, catering to millions of digital "inhabitants," a potential audience of one hundred million iPhone, iPad, and iPod Touch customers, about the same number of Android devices, and hundreds of thousands of other competing handsets. A piece of software that can run on a smartphone can potentially replicate the functionality of an entire piece of hardware or even serve as a replacement for a book, newspaper, or magazine. Thus, rather than invest millions in producing and shipping hardware and physical media, entrepreneurs can now make apps for mobile platforms that compete with entire industries. "Software in theory can be almost anything," Phill said. "It's the closest thing to magic."

Chapter 3

the
always-on
society

The ground shook, the walls rippled, the ceiling exploded, and everything turned dark. Dan Woolley was buried in a mountain of rubble. When he awoke and dug himself out of the pile, he realized his right leg had snapped in half. The back of his head was gushing with blood. The hotel lobby was pitch-black, and to make matters worse, he couldn't find his glasses.

He knew he didn't have long.

First, he needed light. Woolley, a web programmer, came up with a clever idea: He pulled out his digital camera and shined its focusing light on his surroundings. He snapped photos of the wreckage, using the flash to help him search for refuge. Finally— shot after shot after shot—his viewfinder revealed an elevator shaft. He crawled inside and thanked God, but his temporary rush of euphoria quickly sunk. Through his camera he could see the nasty gash in the back of his leg. He was losing blood fast.

Like anyone who has watched movies or a little bit of TV, Woolley knew he had to apply pressure to his wounds to stop the bleeding. But he had no first aid kit and no medical training to do it properly. Fortunately, however, Woolley had another tool. He reached into his pocket and yanked out his iPhone. There was no cellular signal to call for help, but he had downloaded a software application on the phone called Pocket First Aid. He launched the app and tapped the button "Bleeding You Can See." Immediately the phone launched a list of instructions:

Use dressing to stop bleeding with pressure and keep the wound clean.

A dressing can be a gauze pad or any other clean piece of cloth. Choose a size that covers the wound.

Put firm pressure on the dressing over the bleeding area with the flat part of your fingers or the palm of your hand.

You have to press harder to stop severe bleeding. Do not remove the dressing.

Woolley followed the steps carefully. The cleanest articles of clothing he had were his button-down shirt and a sock. Accordingly, he pressed his sock firmly on the back of his head for several minutes and then tied it around. Then he did the same with his leg using his shirt.

The app had one last instruction: *Check for signs of shock. A victim in shock may feel weak, faint or dizzy; act restless, agitated or confused; or be cold and clammy to the touch. Stay with the victim until help arrives.* Woolley mentally check-marked each of the shock symptoms. Since *he* was the victim, he took the last step to mean he should try to resist falling asleep. He set his iPhone alarm to go off every twenty minutes, and he plugged in a battery extender accessory to keep the gadget running.

And then he waited—to be saved or to die.

Realizing the possibility of the latter, Woolley pulled out his Moleskine notepad and a pen. He shined his iPhone light on the paper with his left hand while scribbling last words to his two sons with the right, punctuating the pages with smears of blood:

I was in a big accident. Don't be upset at God. He always provides for his children, even in hard times. I'm still praying that God will get me out, but He may not. But He will always take care of you.

A day passed, and then another day. Lying down in the dark elevator shaft, Woolley used his iPhone to record voice memos to his wife and sons, and he listened to music when he felt discouraged. When the iPhone's battery meter dropped down to 20 percent, Woolley shut it off to save the power. He had stored text messages to send in case he could get a miraculous connection when nearing death.

Then, finally, after a painful sixty-five hours, Woolley heard the scrapes of shovels. A beam of sunlight hit his face. A French rescue team found him alive in the crumbled elevator shaft and pulled him out to safety.

Woolley was one of twenty survivors buried in Montana Hotel in Port-au-Prince when the 2010 Haiti Earthquake struck. His friend David, a filmmaker with whom he was shooting a video about poverty-stricken Haitians, didn't make it. The 7.0-magnitude quake claimed two hundred thousand lives that day, but Woolley ultimately survived with the help of his iPhone, his camera, and a whole lot of luck.

Understandably, Woolley's story of survival made rounds all over newspapers, radio stations, TV networks, and the web. Even more important, Woolley told me, was that the iPhone was an item he carried with him everywhere. "Some people on the web pointed out I should've had a pocket first aid kit," he said. "The reason they're wrong is I wouldn't have it in my pocket." How many people have gone out of their way to add one more thing to their pocket? "What was valuable about the iPhone is it was already in my pocket," he continued. "I thought it would probably be a good way to have some first aid tips in here, so I downloaded that app."

Woolley's incident highlights an incredible social implication of the iPhone and other similar smartphones: an Internet-enabled device, with access to a wealth of apps offering a multitude of utilities, can potentially transform a person into an always-connected, all-knowing being. In Woolley's case, an iPhone app turned him into an amateur medic that helped him survive a natural disaster.

In ways such as this, the iPhone and similar app-powered, always-on smartphones are changing people's lives everywhere in both mundane and magnificent ways. Shopaholics can use their smartphones to scan the barcode of an item they want and immediately find cheaper deals online. Commuters can download an app to see precisely when their next train is arriving. Foodies can point their iPhone's camera at a restaurant and see user reviews of the eatery, the price range of its meals, and what time it closes shop.

Even more inventive are the wealth of niche applications available for smartphones. Military snipers can calculate rifle ballistics with a special app, and soldiers are testing a location-based app that promises to track friends and foes in the heat of battle. Doctors can carry all the medical charts of their patients inside an iPhone app, which they can also use to send prescription requests to pharmacies. The Centers for Disease Control is anticipating outbreaks by crowdsourcing diseases with a GPS-based app.

Anything-anytime-anywhere is unlocking an endlessly expanding list of capabilities, and this is only the beginning. Just imagine the possibilities when the incredibly capable Internet-powered devices we carry in our pockets increase in power and decrease in price while cellular networks mature to handle massive amounts of data at blazingly fast speeds. The potential for humanity to benefit from this revelation is huge, and no prior technology parallels it. Data are entering our thought stream so intimately that the way we engage with the physical world is being significantly enhanced. Every facet of the society we know today will transform as a growing population plugs into the always-on lifestyle.

Most fascinating are the changes the anything-anytime-anywhere will soon deliver to higher education, medicine, law enforcement, and personal broadcasting. Classroom learning will be a richer and more interactive experience than ever. Medicine will become a mobilized profession; anything-anytime-anywhere will even open doors to technology so that we can be our own doctors. Law enforcement officers will be able to carry with them all the resources and tools that are normally tied to police stations. Furthermore, no crime will be left unreported: armed with a camera-equipped smartphone and live streaming-video software, every citizen will have the power to broadcast anything to the world in real time, thus creating a collectively omniscient society of watching eyes.

Education Rebooted

If we can access any information anytime, anywhere in the world with a smartphone, then the way we learn in a classroom is due for an upgrade. Young, bright students (and, heck, the dim-witted ones too) are fully aware that the Internet opens a portal to a live stream of information that billions of minds crowd-source. We have to wonder, how long can the traditional university classroom, in which an instructor assigns a textbook, possibly last? That setting is stale if you look at it objectively: in a classroom, you listen to a single source talk about a printed textbook that will inevitably be outdated in a few years. True, that source is bound to be credible and informed (one would hope), but who's to say there aren't also smart individuals sharing valuable information on the web? There's a clear opportunity here for anything-anytime-anywhere to refresh classroom learning in higher education.

What if, instead of teachers telling students to turn off their cell phones during lecture, they did the opposite? What if universities instead handed out iPhones to use as learning tools inside and outside the classroom?

Abilene Christian University was the first college to try that. Located in Texas, in 2008 the private university began treating every new freshman student and faculty member with a free iPhone. The goal of this technology initiative was to explore how the always-connected iPhone might revolutionize the classroom experience with a dash of digital interactivity.[1]

However, handing out free iPhones wasn't a magic bullet that automatically improved education. The problem with the Internet, after all, is with so much information, determining which data is valuable is difficult. Thus, Abilene Christian had to rethink how teachers lead lectures. Instead of standing in front of a classroom and talking for an hour, Abilene Christian professors act as a guide for the lectures. Professors pick the topic of the day and then instruct students to take out their iPhones and research on the fly. The students then share what they found, and the professor can help assess whether it's good or bad information. With this tweak in approach, suddenly the classroom becomes an interactive lesson in exchanging, discussing, and analyzing information from multiple channels rather than receiving and assessing knowledge from a single source.

This method is like a mash-up of a 1960s teach-in with smartphone technology from the 2000s.

In addition to handing out iPhones, Abilene Christian coded a suite of web apps for students to turn in homework, look up campus maps, watch lecture podcasts, and check class schedules and grades. And just like that, the iPhone is helping Abilene Christian give students the information they need—when and wherever they want it. Didn't understand something during a lecture? Listen to the audio podcast on the iPhone. Want to ask the professor a question? Text message him. Need to look up a library book? Search for it with the campus web app. Can't find the lecture hall where you're taking a final? Launch the campus map app. "It's kind of the TiVoing of education," Abilene Christian professor Bill Rankin told me. "I watch it when I need it and in ways that I need it. And that makes a huge difference."[2]

"About five years ago my students stopped taking notes," he added. "I asked, 'Why are you not taking notes?' And they said, 'Why would we take notes on that? . . . I can go to Wikipedia or go to Google, and I can get all the information I need.'"

Polling—Abilene Christian's new approach to innovate classroom participation—makes their changing approach even more interesting. In some classrooms professors project discussion questions on-screen in a PowerPoint presentation. With polling software that Abilene Christian coded for the iPhone, students can then answer questions anonymously by sending responses electronically with their iPhones. The university's software can also quickly quiz students to help them gauge whether they understand the lesson. That relieves them of any social pressure to appear intelligent (i.e., fear of looking idiotic) in front of their peers. If they answer wrong, nobody will know who gave the wrong answer. And if students don't understand a lesson, they can ask the teacher to repeat it by simply tapping a button on the iPhone.

The idea of polling sounds awfully antisocial, but the Abilene Christian students I spoke to said they are loving it. They say it actually gets more people to participate than the usual handful of sycophants who raise their hands to answer every question asked in the classroom. "Polling opens up new realms for people for discussion," ACU sophomore Tyler Sutphen told me. "It's a lot more interactive for those who aren't as willing to jump up and throw out their answer in class. Instead, you push a button."[3]

Anything-anytime-anywhere transformed Abilene Christian University into a massive data network that is swirling with information from both old and young minds, connecting students and teachers in a more profound way than ever before. Because they are all using the same technology and software, the digital extensions of their minds are completely linked together. To call this a think tank would be imprecise: it's essentially an enormous think *engine* robustly engineered with quality data. Imagine the incredible output of this machine. Now imagine how much an iPhone program, if adopted nationwide, could expedite research, innovation, and scientific discovery on a much larger scale. The implications are truly enormous.

It's also worth noting that in one controlled experiment, Abilene Christian found that students with iPhones slightly outperformed students who took the same courses without iPhones. If executed intelligently, anything-anytime-anywhere can make us more well rounded, sharp, and intelligent than ever before.

"For us, it isn't primarily about the device," Rankin said. "This is a question of, how do we live and learn in the twenty-first century now that we have these sorts of connections? I think this is the next platform for education."

Of course, I'd be naive to predict widespread iPhone adoption will occur rapidly at universities worldwide. Doing so isn't going to be cheap. For Abilene Christian, implementing the iPhone program was neither easy nor inexpensive. In addition to writing custom web apps for the iPhone, the university optimized its campus-wide Wi-Fi to support the thirty-one hundred iPhones currently in use. Rankin, who helped plan the iPhone program during its pilot stage, declined to disclose exact figures for money invested in the iPhone program, but he did state that the initiative only comprises about 1 percent of the university's annual budget. To offset costs, the university discontinued in-dorm computer labs because the vast majority of students already own notebooks. Furthermore, students who opted for iPhones are responsible for paying their own monthly plans with AT&T. Regardless of these cost-saving measures, Abilene Christian is a private university, so the school can afford to splurge on new technology programs occasionally.

Larger public colleges already struggle to provide textbooks and basic resources to their students, so they will be slow to even consider adopting a full-blown iPhone program anytime soon. But there's hope. Competition and the decreasing cost of component parts will inevitably continue to drive the price of smartphones down. Already, the cheapest iPhone costs only $100, and analysts expect that the iPhone will eventually be free with a cellular contract. Perhaps this magical number zero will attract universities to consider such programs. What remains, then, is the issue of smartphone contracts, which will burden frugal students with monthly bills of at least $80 per month. However, the monthly cost of a smartphone might come down as telecom carriers—especially AT&T and Verizon—battle to undercut each other's prices.

Regardless, because our social situation has changed, a smartphone program is a necessary upgrade for colleges worldwide. We no longer need to receive information—after all, data is everywhere—instead, we need to learn and teach one another how to analyze it. By the time our children growing up during this anything-anytime-anywhere revolution become our new generation of teachers, we'll probably see more programs similar to Abilene Christian's sprout up all over the world.

The Doctor on Call

Haiti quake victim Dan Woolley's story of survival with the help of an iPhone app is only a prelude to the host of health benefits that anything-anytime-anywhere is aiming to unlock. Although no single piece of technology will be a panacea for health care issues worldwide, app-powered smartphones are nonetheless gaining serious ground. Researchers and programmers I spoke with are eager to deliver apps and always-on gadgets that make health care and knowledge more easily accessible for everyone.

Together, these developments provide a glimpse into the future of the mobile doctor.

UC Berkeley researchers have developed an accessory called the Cell Scope to enable anyone, anywhere in the world to diagnose malaria and tuberculosis. By attaching the accessory to an iPhone, or any cell phone equipped with a decent camera, anyone can immediately have the ability to perform a diagnosis. A person would take someone's blood or tissue sample and put it on a slide, which you'd then attach to the Cell Scope. A ring of bright LEDs illuminates the sample, and if faint blue dots appear on the image, the patient is positive for malaria. The user can then upload the image to medical experts for analysis and recommendations for treatment.[4]

The Cell Scope's technology is extremely simple, and yet it's a large step toward addressing a substantial issue facing developing countries. Not only do developing countries lack the resources to treat many diseases and health problems, but they also don't have the doctors or the institutions to perform proper diagnoses. However, by giving anyone—not even necessarily a doctor—a Cell Scope to attach to their cell phone, they become immediately capable of breaking down that first barrier of receiving a diagnosis. "The implications are really big, especially for global health and telemedicine," Wilbur Lam, a postdoctoral fellow in bioengineering who is part of the Cell Scope research team, told me. "We're introducing diagnostic capabilities in places that don't have them right now."[5]

There's also a host of apps that doctors are using to carry their work with them everywhere. Epocrates, for example, is an app with access to a massive drug-interaction site, thus giving doctors the ability to view drug information regardless of their location. Doctors also have access to high-end apps to help them keep track of their patients' health chart, such as the iPhone app iChart, which stores everything from patient data to charts and lists of medications, all in a streamlined, organized fashion. Furthermore, to ensure that diagnostic procedures aren't outdated, iChart and Epocrates regularly update with new medical data pulled from health care networks. Farther down the road, doctors hope the iPhone (and, now, Apple's iPad tablet) can completely make obsolete the old-fashioned clipboard and X-ray light box once and for all.[6]

And, of course, there are apps and gadgets aiming to allow us to monitor and treat ourselves anywhere, anytime. In addition to a wealth of first aid apps like the one Woolley used in Haiti, there are apps to test your hearing, monitor your blood pressure and heart rate, and keep track of your weight and diet. There's even an app (that many are uncomfortable to share) called ShyBladder, which helps those who have trouble getting things started in the restroom by playing three different sounds of running water.[7]

In fact, more than any smartphone on the market, the iPhone is equipped to be a medical device. The operating system has a feature that enables apps to communicate with special-purpose accessories, and this opens up a ton of possibilities for the medical field. A great example is Johnson & Johnson's app-powered accessory for diabetes patients, LifeScan. Whereas checking blood sugar levels can be a painful, laborious process for diabetes patients, LifeScan streamlines the process. A specially modified blood glucose monitor hooks up to the iPhone. When a diabetes patient pricks his or her finger, the monitor makes a reading and uploads the data to the iPhone. The LifeScan iPhone app thus serves as a glucose reader that can then calculate a meal builder and an insulin injection schedule. The app can even estimate, based on diet, how much insulin is needed after each meal.[8]

The LifeScan app can also be programmed to alert diabetes patients when they need to make an injection. Before, all of these calculations and measurements would have to be done in a person's head. What's more, the app then lets users send their glucose readings and a message about how they're feeling to their parents, children, or physician.

However, the CellScope, LifeScan, and many similar medical apps have yet to hit the market. These types of technologies take a bit longer because they must be rigorously tested to gain approval from the Food and Drug Administration, among other certifications. But far more intriguing is where the always-on medical technology will take us in the more distant future. For example, a team of scientists, eye surgeons, professors, and students at University of Washington are already working on what will be the culmination of anything-anytime-anywhere for the medical field: a digital contact lens that provides personal health monitoring in real time.[9] With such data, we could be fully aware of our bodily conditions and treat ourselves before illnesses get serious.

The contact lens contains one built-in LED, powered wirelessly with radio frequency, that draws its computing powers from a smartphone. Why a contact lens? Because the surface of the eye contains enough data about the body to make a basic reading of your vital signs, such as cholesterol, sodium, potassium, and glucose levels. Other than that, over a hundred million people already wear contacts, so clearly this would be a huge market. "The eye is our little door into the body," Babak Parviz, a University of Washington professor of bionanotechnology involved in the project, told me.[10]

Because this type of real-time health monitoring has been impossible in the past, there's likely more about the human eye we haven't yet discovered, Parviz said. He and his colleagues have been working on their multipurpose lens since 2004. They integrated miniature antennas, control circuits, an LED, and radio chips into the lens using optoelectronic components they built from scratch. They hope these components will eventually include hundreds of LEDs to display images in front of the eye, such as words, charts, and even photographs. Eventually, more advanced versions of the lens could provide a wealth of information, such as virtual captions scrolling beneath every person or object you see. Parviz's dream? To create a contact lens that will give any wearer the eyes of James Cameron's Terminator. With data everywhere, this fingertip-sized gadget could one day create a new interface for gaming, social networking, and interacting with reality in general.

Sounds neat, doesn't it? However, the group faces a number of challenges before their vision can be actualized. Safety is a primary concern with a device that comes in contact with the eye. To ensure the lens is safe to wear, the group has been testing prototypes on live rabbits, which have successfully worn the lenses for twenty minutes at a time with no adverse effects. Nonetheless, the lens must undergo even more testing before gaining approval from the FDA.

The obvious and most practical implication of always-on health monitoring is that it could potentially eliminate unnecessary trips to the doctor and perhaps even prevent unnecessary treatments doctors prescribe to rack up the bill. (The United States does, after all, burn hundreds of billions of dollars' worth of health care on unnecessary treatments.)

As fascinating, and perhaps incredible, as all these promises sound, however, they won't address the grave issue facing people in developing countries who can't afford such technology to treat themselves or their loved ones. Because of that, we'll likely see always-on medical technology open doors to new medical professions, such as the airborne physician who has his entire clinic in his pocket or the traveling diagnostician who doesn't even need a medical degree.

The Always-On Cop

In January 2009 Virginia woman Rose Maltais picked up her granddaughter Natalie for what was supposed to be a weekend visit. Rose allegedly told Natalie's legal guardians that they would never see her again, and then she drove off with the nine-year-old, according to multiple news reports.[11] Fortunately, tracking down the grandmother was fairly easy because Natalie was carrying a smartphone. Unbeknownst to her grandmother, the Federal Communications Commission has required cell phone carriers since 2005 to provide a way for police to trace phones,[12] and GPS is one of the key tools—a feature on Natalie's device. Police officers called the child's cell phone provider, which was able to triangulate GPS coordinates every time Natalie's phone was activated. Police then entered the coordinates in Google Maps and quickly found the Budget Inn motel where the two missing persons were staying. Rose was arrested after only one night.

Sci-fi movies like *Time Cop* or *Minority Report* foresee a future in which police are equipped with mind-blowing technology, such as time travel, to nab crooks. Realistically, however, these movies make law enforcement look a lot harder than it's actually going to be. In addition to using GPS to help track down victims and suspects, some police departments are already creating special apps for the iPhone in order to ease their jobs. What's more, police officers I spoke with are excited about the impact of anything-anytime-anywhere on law enforcement and the art of crime solving in the even further future. They envision a time where smartphone apps could potentially replace most of the facilities of a police station.

Police officers in Tasmania, Australia, created their own app (not publicly available, of course) to perform car registration checks, thus effectively replacing Radio Dispatch Services personnel. The app uses the iPhone camera to snap a photo of a vehicle, and within seconds it relays the information back to a database of unregistered vehicles and unlicensed and disqualified drivers. The app can also search for outstanding warrants and other offenses. The app then transmits GPS coordinates of where the photo was taken so, if necessary, nearby officers can be alerted to arrive at the scene. Just like that—with the snap of a photo—an officer, even when off-duty and taking his morning walk, can instantly identify and report a lawbreaker.

Furthermore, thieves' practice of stealing people's cell phones is becoming a pretty bad idea, thanks to GPS. In one case in 2009 a Pittsburgh man, after being held at gunpoint and mugged, was able to track down his attackers by loading a website (Apple's MobileMe.com) that provided the precise geographical coordinates of his iPhone on a map. Police followed a blue dot on the real-time map and successfully chased down the iPhone to capture the thieves. Numerous other GPS-related arrests have appeared in headlines in recent years. (In another case last year, a more audacious geek named Kevin Miller tracked down his iPhone thief by himself and eventually confronted the thug. Fortunately, he avoided physical altercation.) Smartphones with GPS could also allow officers to keep each other safe. If each officer were assigned a specific iPhone, and if an officer were in a serious situation and couldn't dispatch his location over the air, the dispatch center could ping his iPhone and get his exact GPS coordinates and direct other officers to him.

Just imagine how much more powerful anything-anytime-anywhere will be as increasingly more departments experiment with apps. A San Francisco police officer, who wished to remain anonymous because of department policies, received an iPhone for work and dreamed up a number of possibilities for apps that could change law enforcement. "It'll change everything," he said. "It'd be amazing if we could scan a detained subject's fingerprint or eyes via the iPhone to access databases to determine who they are. That's a nightly battle and frustrating, because we have to take them to the station to ID them."[13]

Also, given pictures and measurements, an app could potentially turn the iPhone into a tool to recreate a crime scene, the officer added. Such an app would ideally stitch together these elements and upload them straight to a police station's database. He also dreamed of a future in which locks for major businesses were electronic and a special police iPhone app could unlock them in the case of a reported crime or emergency. Cities typically keep key boxes for police officers and firefighters to gain access to buildings, but, he said, they're dated and often unusable.

It all sounds awfully *Minority Report*, and as you probably know, *Minority Report* was hardly a movie about technological utopia. The iPhone, or any smartphone, could become the easiest way for police to invade your privacy.

Text message histories, photos, website histories, e-mails, GPS coordinates, and, maybe later, remote digital fingerprint and eyeball scans—these all add up to suggest that digital data will play an even stronger role in law enforcement in the anything-anytime-anywhere future.

The World Is Watching

"I'm on the top floor of a building in the Fatemi Square and can see they are beating people up with metal and electric batons and you can even hear sound of gunshots," Iranian citizen Fershteh Ghazi tweeted in Farsi. "Slogans 'God is great' and 'death to dictator' are echoing in Tehran," Ghazi posted later. "People are demonstrating on the rooftops."[14]

Dubbed the "Twitter Revolution," the 2009 Iranian election protest took the idea of citizen journalism to new heights. When the government handed President Mahmoud Ahmadinejad an improbable, lopsided victory, many citizens protested, alleging that the government had rigged the election. Riots and street demonstrations then flooded the capital. In an attempt to choke off the flow of news coming from the capital, the government shut down text messaging and phone communication. But these oppressive efforts were no match against the computer-literate populace of Iran, who posted videos, photos, and tweets with their notebooks and smartphones in order to band together for protests and show the world their discontent.

The election protests reached media organizations all over the world, and they then wove together stories from a combination of live blogs, Twitter streams, and visuals posted online. The protests resulted in four thousand arrested and seventy killed—a tragic event for millions of angry citizens.[15] Social media platforms and the web empowered the voices of a population largely suppressed by its government and government-controlled media.

So what's next? Three words: streaming live video. A host of new apps can now turn the iPhone into a video camera capable of live streaming to computers and other phones. The startup

Ustream is leading this new space. In August 2009 the company launched the first app to broadcast live streaming video from your iPhone to the web. That's a magnificent feat because live, anywhere broadcasting used to be a privilege reserved for TV reporters with access to satellite trucks. Now, however, every fleeting, serendipitous, or breaking moment can potentially be documented in real time. We'll have a world of broadcasting eyes. "People always have a cell phone on them," John Ham, cofounder of Ustream, told me. "You can't always predict life, and there are going to be moments you want to share. We've seen people take out devices and streaming earthquakes or planes landing, and there are going to be all sorts of citizen journalism events now if we have millions with this application over iPhone."[16]

The implications are hugely beneficial. When billions of people own smartphones with live video software and telecom networks in turn speed up to handle the insane bandwidth required for ubiquitous broadcasting, we will live in a world in which everyone is holding each other accountable. Committing any sort of wrongdoing, such as stealing, polluting, littering, child abuse, or sexual harassment, will be more difficult. We will be able to effectively coexist with one another at all times. We can share our kids' soccer games, our weddings, the concerts we attend, or the lectures we're sitting through. Soon, we will have video conferencing phones just like the Jetsons; information communication and exchange will be easier and more lucid than ever.

But wait a minute—there will be video cameras everywhere. Sounds familiar, doesn't it? The UK in the 1990s deployed its controversial closed-circuit television system—roughly 1.5 million surveillance cameras in city centers, stations, airports, and major retail areas—in an effort to deter crime. But critics have provided considerable evidence that the CCTV system has been largely ineffective in preventing crime; instead, the UK's program is often criticized as a massive invasion of privacy.[17]

However, the fundamental flaw with CCTV is that it's a closed system, so the live video is only broadcasted to a limited set of monitors, which a limited number of eyes watch. The difference with live streaming video on smartphones is subtle but significant: you broadcast to an open channel that anyone can view on the web. That gives anybody the ability to engage in the moment and influence the outcome if they so wish. For example, Ham said one of the most exciting uses of Ustream took place in the weeks leading up to the presidential election. A family had installed a Ustream camera in front of an Obama sign to see if anyone would try to steal it. Sure enough, someone did, and people watching it on Ustream were able to discuss the event while it was happening, with one person even going so far as to call the police. Police actually arrived in time to reprimand the individual.

The live video phenomenon is only beginning to catch on, and it's poised for rapid growth in the next few years. In January 2010 Ustream received $75 million from Softbank, a Japanese telecom company, to expand live streaming infrastructure in Asia. Ham also plans to use the funding to optimize Ustream's resources in the rest of the world.

All in all, more eyes watching and more minds participating amount to a more connected, transparent society. Just like Apple depended on the genius of crowdsourcing to churn out iPhone apps, the era of live video will rely on the collective wisdom and diligence of the masses to paint an accurate picture of the world's events. In turn, our role as media viewers is bound to increase in importance. We won't simply consume media anymore; rather, we'll experience it and be able to influence what's actually happening. "This comes down to a fully transparent society," Ham said. "Nothing is faster than live video in terms of delivering the most information in the quickest amount of time."

The Always-On Society

The iPhone is empowering and connecting people in a profound, unprecedented way. Apple accomplished this feat by significantly improving both quality and quantity of data. Furthermore, iPhone apps fit the human way of working: they employ clean interfaces that present useful data—tools designed to accomplish specific tasks. What's more, there are enough apps to fill every practical need and every niche, thus making the iPhone so customizable that it can be tailored to suit any lifestyle and profession. Hundreds of millions of people now own app-powered smartphones that have radically changed their lives, thereby leveraging their abilities to do their jobs and to engage with the real world. Health care, law enforcement, and education are only a few key examples of professions poised for transformation in the anything-anytime-anywhere future. Everything is going to change.

Chapter 4

skyscraper businesses

Tucked in the corner of San Francisco's Mission and Embarcadero streets, Boulevard restaurant is one of the most exquisite eateries in the country. Nancy Oakes, the city's top-rated chef, owns the restaurant, and her dishes include a wood oven–roasted pork prime rib chop, pan-roasted Carolina wreckfish, and Sonoma foie gras. Patrons pay about $200 a head for Boulevard's fine dining experience.

Also part of the Boulevard experience is that you can make a reservation for the restaurant anytime from anywhere instantly by using a computer or smartphone. Rather than call to ask about the next open spot, you can view all the available time slots in a digital calendar, set the number of people in your party, hit the "Book" button, and you're ready to eat. This all happens in real time, meaning there are no reservation overlaps between diners; it's a headache-free process.

The system that makes this anything-anytime-anywhere booking experience possible is a thirteen-year-old company called OpenTable. The most successful online-reservation service in the world, OpenTable handles Internet booking for fifteen thousand of the thirty-five thousand restaurants that use online reservations across the United States, and in the third quarter of 2010 they seated 15.4 million diners. The company's market capitalization in March 2011 was about $2 billion."[1] That's a lot of money and a huge chunk—about 43 percent—of restaurants that use online reservation sites have opted to use OpenTable. And to succeed, OpenTable's strategy, it turns out, wasn't open at all.

Contrary to its company name, OpenTable flourished because it actually operates as a closed network. For a restaurant to participate, it has to use OpenTable's provided computer modules and proprietary booking software, which connects with OpenTable's exclusive online database of diners. That racks up to a number of costs for a restaurateur: a one-time sign-up fee of about $650 to set up and then an average of $270 per month for the hardware and table-management software (sometimes more if the restaurant has multiple modules). Finally, each time a diner books a reservation through OpenTable's website, the restaurant is charged $1. If the diner books a reservation with OpenTable through the restaurant's own website, then the restaurant is charged 25 cents. OpenTable has a business model of pay-to-play, and restaurants that opt against using the service don't have access to that database of millions of OpenTable diners.[2] Therefore, any restaurant owner who wants to lure in this brave new world of always-on diners must debate whether it can succeed with traditional phone reservations and word-of-mouth recommendations or if paying thousands of dollars each year to an Internet service to bring in some extra customers is worth it.

The benefits of OpenTable are big on multiple levels. Customers get to book reservations in a manner that's instantaneous and convenient. Restaurant owners get their restaurants exposed to potentially millions of customers. Finally, restaurant hosts spend less time picking up the phone and can use OpenTable's software to seat people at tables, whereas before they had to pencil in seating arrangements on sheets of paper.

The drawbacks are less obvious. After closely analyzing the numbers, the real cost for each reservation booked through OpenTable can be a lot stiffer. For example, say a party booked through OpenTable pays $40 for a meal. $1 of that $40 goes to OpenTable, which is 2.5 percent. That number is significant when you take into consideration that the restaurant industry's average profit margin is 5 percent.

Furthermore, if we look even more closely at OpenTable, the price a restaurant owner pays may be a lot more than a buck. Looking at OpenTable's SEC filings, technology marketing guru Jonathan Wegener found that 57 percent of OpenTable diners book directly through OpenTable.com and the remaining 43 percent book with the OpenTable tool at the restaurant's own website.[3] The average restaurant spends $515 on OpenTable and gets 345 diners each month from the service. Although that looks good at first glance, we must keep in mind that 43 percent of those customers booked through the restaurant's website, meaning that if they didn't use the OpenTable.com directory to find it, they already chose to eat at that restaurant. The real value are those 57 percent, 197 customers, booking through OpenTable.com who may have never known about the restaurant. Therefore, restaurants are paying $515 to gain 197 new customers, so the total comes to $2.61 per customer, Wegener argues.[4]

So for a party of four, that would come out to $10.40 that the restaurant pays OpenTable to bring in new customers, according to Wegener's math. Mark Pastore, owner of the restaurant Incanto in San Francisco, who has resisted signing up for OpenTable, says that some restaurant owners who use OpenTable complain that they feel "held hostage" by the hefty fees and the lack of a viable alternative. "OpenTable affects my business either way, but I feel like it costs me less money not to be on them than to be on them," he said.[5]

OpenTable is not a monopoly—it would be if 100 percent of restaurants using online reservations signed up for OpenTable— but because nobody else has a similar grab of the wired-restaurant market share, it can continue to keep fees as high as it wants.

So how, then, did OpenTable get so big whereas other competing companies did not? OpenTable did something right that nobody else quite figured out. Some companies offered tools for restaurants to create their own online reservation systems, and others offered software to help diners find restaurants to eat at. OpenTable, however, rolled both these ideas into one to create an online booking experience. "OpenTable's become a lot more than just a convenient set of tools," Pastore added. "They've become a market force." Each year Pastore has studied OpenTable and resisted using the tool because of the high costs of operation, and many of his fellow restaurateurs complain that whether the monetary gains are worth the costs remains unclear.

OpenTable is a brilliant execution of a closed business model known as vertical integration, in which a single entity seeks to own, command, and control key strategic areas of its business operation, whereas a horizontal business largely relies on outsourcing operations to multiple parties. With a vertically integrated approach, OpenTable is a one-stop shop for a technological solution to make online reservations convenient for two groups: the restaurant workers and the restaurant customers. The OpenTable website, smartphone apps, user database, table-management computers and software come together as one easy-to-use, friction-free experience. That's perfect for restaurant owners, who are far too busy to deal with this type of technology on their own, as well as busy travelers, businessmen, or just about anyone looking to make plans to grab a bite. Creating this seamless experience wouldn't have been possible without vertical integration.

However, one large question lingers: is OpenTable creating new economic value, or is it exploiting the inherent value of restaurants? That is, is a vertically integrated company good just for the businesses, or does everybody win—and if so, by how much?

In the case of OpenTable, do the restaurateurs benefit more, or as much, as OpenTable? Pastore doesn't think so. "Very few restaurateurs sign a $5,000 check to OpenTable and say, 'I'm so thrilled, I got every penny out of that,'" Pastore said. "Most of them feel a little trapped by OpenTable." OpenTable claims it doesn't have figures to prove just how many "incremental" customers it brings to restaurants—that is, new patrons who would never have known about the restaurant without OpenTable. So the answer to that question remains unclear.

The fact that vertical integration has proven time and time again to be very successful is very intriguing. After all, it's frequently associated with the words "closed" and "control," which have negative connotations in a society that boasts democracy, freedom, and choice. Vertical integration is the same strategy that made Apple the most valuable company in the world, a corporation famous for controlling every aspect of its products. Without vertical integration, the iPhone would never have unlocked the anything-anytime-anywhere experience that so many enjoy today. Vertical integration is also the model that made Starbucks the biggest, most influential coffee retailer on the planet—and, conversely, Starbucks' loss of vertical control is what led to its downfall. Along with OpenTable, these corporate giants demonstrate that a vertical strategy wins against more "horizontal" businesses that heavily rely on third parties.

How Starbucks Ground Its Way to the Top

It all started with a bean.

Dave Olsen, former chief coffee procurer of Starbucks, searched the world for the magical bean that would eventually make Starbucks a java superstar. Scouring the mountain trails of Guatemala, Kenya, Indonesia, and other parts of the globe, Olsen would only settle for the *arabica,* a superior bean compared to the cheap *robusta* found in commoditized American coffee at the time. Eventually, by 1992, Olsen homed in on the ultimate opportunity: the sale of the Narino Supremo, a very small crop that grows only in the high regions of the Cordillera mountains—a bean so flavorful and unique that Western Europeans would guard it from outsiders. Starbucks negotiated a deal to purchase the entire yield, and the company became the sole provider of one of the highest-quality coffees in the world.[6]

The next step was the roast. Starbucks created its own facilities to handle roasting and distribution of these coveted coffee beans. Then, for hiring, the company handpicked each roaster and trained her for over a year; becoming a Starbucks roaster was considered an honor. Starbucks' custom-designed roasting hardware consists of a powerful gas-fired machine to blaze the beans; the human roasters then use a combination of computer software and their own senses (sight, smell, and hearing) to determine when the beans are perfectly roasted.

Starbucks cultivates a unique culture as well. The company offers premium benefits packages to both part- and full-time employees that would put many companies to shame: health insurance, dental, 401K savings, stock options, an employee assistance program, extensive training workshops, and even dependent coverage for same-sex partners. These benefits—the stock option, especially—ensure employee loyalty, meaning the company has less turnover and the customer enjoys a more consistent overall Starbucks experience.

Starbucks' vertically integrated approach blended into a unique cafe experience that charmed the five senses. The second that customers stepped into a store, the aroma of fresh, high-quality coffee beans would invade their nostrils. And when they ordered a cup of joe, they could watch the barista scoop the beans out of a bin and then listen to the sounds of the grinder before the coffee dripped slowly into a cup. At the end, customers wrapped their hands around a hot cup of coffee brewed at the perfect temperature before tasting their precious Starbucks concoction. It's no wonder why Starbucks became such a phenomenon.

The rest of the story is about branding, distribution, and retail strategy. In a nutshell, Starbucks went for the approach of being everywhere. In practically any city you could probably be able to find at least three Starbucks locations within a few blocks of one another. The reason this worked was because Starbucks manages its own distribution, so it could more efficiently handle large deliveries to coffee shops that were clustered in proximity to each other. Every shop would also be less crowded because patrons would be spread throughout Starbucks locations in their neighborhoods; there would be more foot traffic going in and out of the stores overall. Unlike most companies, Starbucks didn't fear cannibalism; in fact, the coffee giant thrived on it.

That's Starbucks' success story summed up in four paragraphs, but it took about two decades for Starbucks to identify, execute, and refine its strategy. Similar to OpenTable and Apple, the key to Starbucks' enormous success was a controlled, quality experience that competitors at the time couldn't match. Starbucks took off because its strategy was vertical. It didn't grow its own beans, but it did carefully secure the best partnerships and purchase premium crops to meet quality standards. Altogether, the Starbucks experience added up to a mega sensation, pushing specialty coffee to its tipping point in the United States. Suddenly, coffee was something people would specifically go out for rather than make at home with instant powder and water.

And then came the fall.

Starbucks just kept getting bigger and bigger. From the 1990s to 2000s Starbucks had grown from less than one thousand stores to over thirteen thousand locations; by 2002 its sales had skyrocketed 2,200 percent over the decade.[7] As a result, shops needed to brew a lot more coffee and do it a lot faster. They started using automatic espresso makers, replacing their premium, semiautomatic La Marzocco machines to save some precious time. To serve fresh roasted coffee in every store at every minute of every day, Starbucks employees could no longer waste time scooping fresh coffee out of a bin, grinding it in front of customers, and slowly dripping it into a cup. The company moved toward "flavor locking" coffee packages, which lacked the strong aroma of the fresh beans.

To make a long story short, to get bigger, Starbucks lost control of its core values and killed the Starbucks experience.[8] The company's once-vertical approach became horizontal. Duly, in 2008 Starbucks saw its first sales decline.[9]

By 2009, when the US recession really soured, Starbucks took a nosedive. Profits fell 77 percent.[10] The crumbling coffee titan started trying out some new business recipes. It began positioning itself toward value-conscious customers. In March 2009 Starbucks started offering a breakfast value meal: a cup of coffee plus an egg sandwich for $4. The company even began shutting down stores. This all seemed desperate for a brand that was once viewed as higher class and special.

"We achieved fresh roasted bagged coffee, but at what cost?" lamented Starbucks chairman Howard Schultz in a memo sent to staff when sales began declining. "I take full responsibility myself, but we desperately need to look into the mirror and realize it's time to get back to the core and make the changes necessary to evoke the heritage, the tradition, and the passion that we all have for the true Starbucks experience."[11]

And then it all came back to the beans.

Starbucks decided to go vertical again, except this time more extreme than ever before. In late 2010 it opened its first coffee bean farm and processing plant in China.[12] The chain is hoping to expand its eight hundred stores in China by convincing more Chinese to switch from tea to coffee. In other words, Starbucks wants to do for China what it did for the United States: make coffee mainstream by offering a quality experience. It's like starting over one's life in a new country, humbled by one's mistakes.

Apple's Vertical Comeback

"We build the whole widget."

That's one of Steve Jobs' favorite ways to describe Apple's approach to creating product, and it sums up his company so well.

Apple is the most vertically integrated corporation in the world. From the buttons to power adapters and from computer materials to packaging design, Apple employees oversee every element of an Apple product, and nothing escapes the scrutiny of their CEO. "We don't take off-the-shelf parts, [and add to them] huge, major components from other companies, then throw our operating system on it," said Tom Boger, Apple's senior director of desktops, in a 2006 interview with TG Daily. "We build the whole widget from the ground up. We start with the industrial design, we do all the electrical engineering. Every single aspect about a Mac has been designed by Apple."[13]

Besides the product design, everything about the Apple ecosystem is also extremely vertical. The iTunes Store syncs only with Apple's mobile devices. The Mac operating system runs only on Apple computers. The Apple TV media box plays iTunes videos and songs. When you buy an Apple gadget, you don't simply buy a gadget; instead, you plug into an Apple world that's one-of-a-kind, convenient, and exclusive.

However, that Apple's vertical strategy didn't win from the start is interesting to note. In the early PC days Apple pioneered a graphical user interface for an operating system and then made it exclusively for Apple computers—many other computer makers were using a similarly vertical approach at the time as well. Then a young Bill Gates took a GUI-based software product, Windows, and licensed it to third-party manufacturers, who then sold computers less costly than Apple's. This key turning point ceded dominance to Microsoft during the PC revolution. In recent years Microsoft still consistently holds about 90 percent of the PC OS market share, whereas Apple's Mac only has about 5 percent.[14]

But market share numbers don't tell the entire story. Just because Apple had a small grab of the market share compared to Microsoft doesn't mean it was out of the game; in fact, Apple was still an incredibly successful niche. The beginning of Apple's downfall occurred after Jobs was fired over a power struggle and the company attempted to be horizontal.

Under the leadership of Michael Spindler, Apple started a clone program in the mid-1990s in which the Mac OS was licensed to third-party manufacturers, including Motorola, Power Computer, Umax, and Radius. The clone episode was Apple's biggest nightmare ever. Power Computing consistently shipped Mac clones that turned out to be faster *and* cheaper than Apple's Macs. Power Computing shipped 100,000 clones in its first year in business, which cut directly into Apple's bottom line. As a result, sales of Apple's Power Macintoshes took a big hit. Though Spindler believed the strategy would increase market share, the program actually lost money for Apple and third-party vendors using the OS. In the background of the clone wars, Microsoft in 1995 released a new version of its operating system, Windows 95. Windows PCs were generally cheaper than Macs, and budget-conscious customers voted with their wallets.[15] By 1996 Apple had hit an all-time low and nearly went bankrupt.[16]

Jobs returned to Apple in 1996 as a consultant and in 1997 retook his throne as CEO. He began restoring his company to its original state: a vertically integrated business. One of the first items on his agenda was to end the clone program for good. He announced a deal to buy all Mac-related assets from Power Computing, which had become the biggest Mac cloner on the market, effectively killing the company. Then Apple backed out of extended license deals with other Mac licensees, which ultimately halted the program.[17] Jobs said, "I went to the clone vendors, and I said guys, we're going to go broke doing this. And if we go down the shitter, the whole ecosystem will go down the shitter, and you won't be here either."[18]

In 1998 Jobs introduced the iMac—a machine that included everything you needed in a PC: the CPU, hard drive, Internet networking gear, and more were all crammed inside the plastic monitor. It was a complete system in a box, unlike anything else out there at the time. The iMac was a huge hit, and its sales drove Apple back to profitability.

Later, the iPod would be Apple's next big move into vertical integration. The strategy here was peculiar: music purchased through the iTunes Store, the first major digital music retailer, would work only with the iPod. However, Apple did make the iTunes software available for Windows as well as Mac users. In this particular case, Apple was boldly creating a vertical market even within its rival's territory. The strategy worked: tied to Apple's exclusive digital music store, the iPod would later consume 90 percent of the MP3-player market share.

And then came the iPhone. Apple's philosophy of airtight control came together for this device: Apple designs the iPhone's processor chip, and most importantly, Jobs ensured that Apple had complete control of the software so telecom carriers couldn't cram it with their junk. Furthering his vertical strategy, Jobs opened a digital software storefront, the App Store, which was exclusive to Apple's iPhone, and later the iPod Touch and iPad. Vertical integration enabled Apple to accumulate hundreds of thousands of apps, making its mobile products more attractive than ever. As a result, Apple sold 100 million iPhones over three and a half years, and nearly 15 million iPads in one year. By 2011, the App Store quickly crescendoed into a $2 billion industry.[19]

This story arc is similar to Starbucks'. Starbucks started out as a small, vertically integrated company and built its success from a quality coffee experience. Apple started out in a similar way and thus succeeded as a small company that offered a unique, premium PC experience. Both companies plummeted when they went horizontal. Now Apple is hugely successful and profitable, having reclaimed its vertical model, and perhaps Starbucks will see similar results now that it's growing its own beans in China.

So why was vertical integration a losing strategy during the PC boom, and why is it winning today in the smartphone boom? Horizontal PC makers were able to mass produce powerful computers using standard components, beating vertical manufacturers such as Apple, which sold more expensive machines. However, in the case of the mobile industry, the horizontal approach didn't succeed because prior to the iPhone, telecom carriers largely had control over how phones were made. When working with Microsoft to make Windows Mobile phones, for example, manufacturers issued the software company a set of instructions, and Microsoft engineers would then modify a skin of Windows Mobile to adhere to those instructions for a specific phone model.

Next, there's pricing, which is frequently brought up when discussing Apple. Apple is typically viewed as a luxury brand. Although Apple does, for the most part, target the higher segment of the PC industry—all MacBooks cost at least $1,000, for example—the pricing argument doesn't hold much relevance anymore when discussing the iPhone. The iPhone costs $200 with a two-year contract with AT&T or Verizon, and Apple continues to sell the older-generation iPhones for $100. So why wouldn't Joe Schmoe get an iPhone? The most expensive part of owning an iPhone is the monthly cost—at least $90 per month. However, similar handsets running the Android-powered OS require roughly the same monthly rate for voice and Internet data.

Finally, another argument to consider is that the technology landscape has completely transformed over a few decades. Everybody who wants a PC today already has one, whether it be a Mac, Windows PC, or Linux box, whereas before, computer makers catered their products to business users. "If Apple represents the shiny, happy future of the tech industry, it also looks a lot like our cat-o'-nine-tails past," wrote Leander Kahney, a former news editor at *Wired* and author of the book *Inside Steve's Brain.* "In part, that's because the tech business itself more and more resembles an old-line consumer industry. When hardware and software makers were focused on winning business clients, price and interoperability were more important than the user experience. But now that consumers make up the most profitable market segment, usability and design have become priorities. Customers expect a reliable and intuitive experience—just like they do with any other consumer product."

Thus, technology has become mainstream, and as a result, people's needs have changed. Back in the day when PCs were still young, Microsoft's model of offering a "good enough" computer worked. Today, now that computers are more affordable and accessible, the United States has become an information-driven economy. "Good enough" is no longer good enough. People dream of technology that's instantaneously gratifying, convenient, and caters to their every need. The iPhone was the first gadget to deliver that dream of anything and anytime from anywhere, and a vertically integrated solution was fundamentally necessary to wrestle control away from the carriers and design a phone around customer enjoyment.

A question often raised in the technology industry is whether the iPhone's verticalized strategy has long-term potential. Apple's biggest competitor in the phone market right now is still Google's

Android operating system, a horizontally integrated platform. Google's Android is an open-sourced operating system, which means any manufacturer can freely take the OS, modify it however they wish, and ship it with their own hardware. Several large manufacturers, including Motorola, HTC, and LG, opted to make Android-powered phones, which are available on every telecom carrier in the world. As a result, Android is winning in terms of numbers: There are more smartphones sold running Android than phones running Apple's exclusive iOS. However, keeping these market share numbers in perspective is important.

In the third quarter of 2010, 43.6 percent of smartphones shipped in the United States ran Android, whereas 26.2 percent of smartphones shipped were Apple's iPhone, according to marketing firm Canalys.[20] So that means when you add together multiple phone models from different manufacturers, you have a bigger number of Android phones sold than iPhones. However, Apple's iPhone is the second most popular operating system running on a single piece of hardware made by a single company. When we look at the big picture, Android may be the most dominant platform, but Apple's iPhone is the best-selling piece of hardware. Given that Apple makes most of its bucks primarily on hardware, this is a win for Apple.

Furthermore, the third-party programmers also reap the benefits of vertical integration on the iPhone. Because apps sold through the App Store only work on Apple's mobile devices, independent programmers could easily create apps for a single audience of iPhone, iPod Touch, and iPad customers. The App Store is a substore within the iTunes Store, which already has hundreds of millions of registered customers. Everybody who owns an iPhone, iPod Touch, and iPad has only one way to buy an app: they log in with their iTunes ID and hit "Buy."

By contrast, the competing Android app platform suffers from a fundamental issue of fragmentation. Not every Android app will work on every Android phone, as a different manufacturer makes each kind of device; one phone might have a certain feature that another one doesn't, so the same app might not work the same on two different Android devices. Payment is also an issue of fragmentation on the Android platform. Google doesn't have an established media platform comparable to iTunes for electronic payments. There's Google Checkout, but not every Android user has an account, and some people never bother signing up. So some people will buy an app with Google Checkout, some will manually enter in their credit card information to buy only one app, and others won't buy anything at all. As a result, the majority of popular Android apps are free and include advertisements because Android programmers have concluded that priced apps just won't sell very well. Conversely, a vertically integrated app store gave Apple programmers a cohesive audience, thus making it easier to make money, which is why the vast majority of them chose to make content for Apple, thereby furthering the success of the iPhone.

So will Apple "lose" with vertical integration in the long term like it did against Microsoft during the PC revolution? This is very unlikely, and now that the landscape has changed, vertical integration will likely continue to win for Apple, a company that makes most of its money selling hardware. As of this writing, in the United States the iPhone finally became available on Verizon after three and a half years of being exclusive to AT&T. That Apple's vertical strategy will continue to win seems a safe prediction for a number of reasons: 1) the iPhone costs about the same as an Android phone does monthly, so the price argument is moot; 2) the iPhone App Store has more third-party apps than

Android and will likely continue to do so because the App Store makes more money for developers than Google's platform; 3) the iPhone, as the numbers show, is the most popular piece of hardware.

When it comes to arguing Android versus iPhone, there are plenty of political arguments we can raise about closed-source versus open-source software, but for the most part most customers aren't going to care. The average customer wants an affordable smartphone that works easily and can do a lot. Given the aforementioned reasons, why Joe Schmoe would buy an Android phone instead of the iPhone is difficult to understand. In the mobile industry, vertical integration is going to be the big winner for Apple, its customers, and programmers. We get the whole widget.

Microsoft, meanwhile, is still a highly profitable business, but many would agree that the software giant is losing the post-PC era. Hardly anyone today uses a Windows phone. As a result, Microsoft, too, is trying out vertical integration with its latest phone operating system.

Microsoft's Windows Phone Reboot

"Whatever device you use, now or in the future, Windows will be there."

That's Microsoft's business philosophy in a nutshell, straight from CEO Steve Ballmer's mouth. Microsoft wants to be everywhere. It wants Windows on every computer. It wants everyone using Microsoft Office for work; the joke at Microsoft headquarters is that designing Office is like "making a pizza to feed a billion people." The bottom line is that Microsoft wants everyone in the world to use Microsoft software on every type of computing gadget that exists. This is a horizontal business model, as the items that make the system are divided between separate parties: manufacturers such as Toshiba and Sony make the computers and Microsoft provides the software.

But Microsoft's horizontal strategy didn't work out so well for the Windows phone.

Microsoft staff refer to December 2008 as "The Reset"—the month that the company killed all progress on its Windows phone project and started over. That employees I interviewed were unanimous in calling this a good thing is a measure of how deep a hole Microsoft had dug itself into.

The software titan had a head start on phone software beginning with Windows CE back in 1996, which would later become Windows Mobile. However, when the iPhone and Android phones hit the market, the Windows Mobile OS suffered steep declines in market share.

In the phone market, Microsoft went from being the market leader to becoming nearly irrelevant within the span of a few years. Windows Mobile crumbled for a number of reasons, which can all be traced to its DNA. Before smartphones went mainstream, their designers catered applications to enterprise-oriented "power" users. Like the earliest computers, user experience wasn't a top priority—only interoperability. "It was trying to put too much functionality in front of the user at one time," said Bill Flora, a design director at Microsoft, reflecting on Windows Mobile's mistakes. "It resulted in an experience that was a little cluttered and overwhelming for a lot of people today. It felt 'computery.'"[21]

The cluttered design was the result of making a mobile operating system based on instructions from phone manufacturers and telecom carriers. Because there were so many different types of phones out there, the solution was to cram as many features as possible into a phone screen in order to have the best chance of catering to a broad audience. However, this horizontal integration made the Windows phone unfriendly to the average gadget user.

An unsexy OS didn't bode well for Microsoft. The outdated design of Windows Mobile contributed to a stereotype that Microsoft cared little about customers and was focused only on big sales to big companies. Its design symbolized a software leader losing its edge. Furthermore, Windows Mobile's shrinkage in the market was embarrassing for a company whose CEO, Steve Ballmer, previously laughed at Apple's iPhone for its lack of a keyboard and high price tag, only to admit three years later that Microsoft had fallen far behind. "We were ahead of this game and now we find ourselves No. 5 in the market," Ballmer said at an All Things Digital Conference. "We missed a whole cycle."[22]

So the software giant made a bold decision. Recognizing it needed to play serious catch-up, Microsoft essentially hit CTRL+ALT+DEL on Windows Mobile, rebooting its mobile OS like a balky, old Windows PC and making a fresh start. The company spent six weeks hatching a plan for a Windows phone do-over, setting a deadline of one year to build and ship a brand new OS. The end result was Windows Phone 7, an operating system with a beautiful, tile-based user interface that's far more user-friendly than its predecessor.

The reset, however, was no simple task. Microsoft hired new managers, reorganized the Windows phone-design department, and created new lab facilities to do quality control on the software. Microsoft was now tightly controlling the creation of a brand-new mobile OS; in other words, Microsoft's phone department was going vertical.

This new zeitgeist started with a new leader. Ballmer appointed Microsoft veteran Andy Lees as senior vice president of the Mobile Communications Business in mid-2008, and Lees assessed the phone division's future. Apparently Lees wasn't pleased with where it was going. After consulting with engineers and top managers, Lees made the call to scrap Windows Mobile 7, a project Microsoft had been developing (and delaying) for more than a year.

Shortly after Lees hit the Reset button on the Windows phone project, he recruited Joe Belfiore, at the time leader of the Zune division, to direct, as corporate vice president of Windows Phone, the creation of a brand-new mobile OS. It's worth noting that Zune music player was a vertically integrated product—Microsoft's attempt to compete with Apple's iPod. Microsoft oversaw the Zune's manufacturing design as well as the operating system that powered it, and it opened a Zune music store to compete with iTunes. So why Lees appointed Belfiore seems clear: he would be the man in charge of the Windows Phone's new vertical regime. "We're taking responsibility holistically for the product," Belfiore said. "It's a very human-centric way of thinking about it. A real person is going to pick up a phone in their hand, choose one, buy it, leave the store, configure it and live with it for two years. That's determined by the hardware, software, application and services. We're trying to think about all those parts such that the human experience is great."

In addition to starting with a blank slate with new faces, Microsoft also chose to embrace an entirely different mobile strategy. Windows Mobile's old MO was to modify skins of its OS in compliance with a manufacturer's set of instructions. The new plan for Windows Phone 7 was to design the operating system around customer enjoyment, similar to Apple's approach.

Microsoft would continue its strategy of licensing its OS to manufacturers, but this time it would set the rules: all phones running Windows Phone 7 must meet a criteria of hardware requirements (three physical hardware buttons and a specific CPU, for example). Also, every device must pass a series of lab tests conducted by robots that Microsoft engineers designed. These stringent requirements aim to ensure that Windows Phone 7 works consistently across different devices, Belfiore said. If a manufacturer's phone failed a test, a Microsoft engineer would send the phone back and tell the manufacturer how to fix the problem. "The team psychology was, 'Here's an OEM saying we want to sell a million phones,'" said Belfiore, reflecting on the previous mobile strategy. (OEMs are "original equipment manufacturers," the makers of computers and phones who have been the backbone of Microsoft's customer base for over three decades.) "The primary customer was the OEM. Now the target is the person [who owns the phone]."

Belfiore wasn't shy about criticizing Google's Android OS. Even though Google currently dominates the mobile OS market, its strategy of licensing the Android OS to manufacturers is similar to Microsoft's previous approach with Windows Mobile: it's open-ended with few restrictions on how manufacturers can use or modify the OS. As a result, Android is suffering from some of the same issues as Windows Mobile did: Android works better on some phones than others, manufacturers are shipping different versions of the OS on different phones, some Android phones are shipping with bloatware made by carriers,* and some app developers complain that making software is difficult

---

*A term for software that comes preloaded on hardware; often the software tends to be junk that slows performance.

because of the hardware and OS fragmentation. Belfiore said Microsoft's new mobile strategy would exert control over the OS so that it's one tidy platform, making it easy for customers to know what they get when they buy a Windows Phone 7 device while allowing third-party programmers to make apps for multiple devices with zero headaches.

He added that bloatware would not be an issue because Microsoft has reached an agreement with manufacturers and mobile operators. Phones will ship with half of the front screen reserved for carriers' and manufacturers' custom apps (the Samsung Focus phone, for example, includes an AT&T GPS app), and Microsoft gets the rest of the screen for its default apps (e-mail, calendar, address book, etc.). Furthermore, any of these apps can be removed from the front screen if customers don't enjoy them.

However, a cleaner, consistent user interface wasn't achieved just by tweaking mobile strategy; Microsoft had to verticalize Windows Phone 7's design as well. Microsoft's design director Flora established "Metro," a set of design standards to guide engineers designing the new operating system. "The philosophy of Metro is trying to do a lot with a little," Flora said. "Use typography in a creative way and eliminate a lot of decoration. Just have the form of it tell the story. That allows the content really to be the hero."[23]

Thus, instead of a screen cluttered with icon buttons, Windows Phone 7 focuses heavily on typography to display different features and functions. What's more, the Home screen of Windows Phone 7 is a set of large tiles that users can customize to suit their preferences.

Metro is based on decades of design principles that Microsoft created and iterated upon going as far back as its Encarta Encyclopedia in 1995 as well as taking the best design philosophies from products such as Xbox, Windows, and Zune, Flora explained. He said he was "evangelizing" the Metro style throughout the company's many divisions. "Metro allows the different brands in Microsoft to be their own way, but with a common and consistent thread that holds them together," he said.

In addition to adopting a set of design standards, Microsoft also completely reorganized the Windows Phone design department. Albert Shum, who previously designed watches and exercise gadgets at Nike (including Nike+), spearheaded the reorg as director of Microsoft's mobile-design team. At Microsoft Shum manages a team of sixty designers in an open office (i.e., no cubicles), and he divides the designers into two groups: left- and right-brained thinkers. The left-brainers handled the hard-core engineering work, whereas the right-brainers focused on interface, and then the groups met regularly to discuss the project.

Shum compared managing the design of Windows Phone 7 to being a movie director. "People have the script but still need a director to drive the process along," he said. "Software is like making a movie and building a skyscraper. You're not quite sure how it's going to stand until it comes out in the end."

"Here's our new movie," he added. "Hopefully you'll like it."

Building a skyscraper is truly a brilliant way to describe the creation of Windows Phone 7. The company completely started over and built the new OS from the ground up with vertical integration: everything from design to QA testing, from the staff organization to the manufacturing strategy, was verticalized.

It's important to note, however, that Microsoft's core business strategy is still horizontal: Windows Phone 7 still involves working with third-party manufacturers that make the phone hardware. But with the new vertical process, Microsoft surrenders zero control over the OS, and with its strategy it even exerts control over its partners. Microsoft has vertically integrated parts of its phone-creation process that actually matter, thereby taking control of its mobile-software destiny.

A Vertical Future

The common thread between these vertical businesses is that they are not product designers; rather, they are experience designers. Apple's iPod has access to an Apple music store to get content; the iPhone has an Apple app store to expand its capabilities with additional software; the Mac runs Apple's exclusive Mac operating system on Apple's hardware that was built in-house. Likewise, Starbucks became a global sensation with not only quality beans but also a tightly controlled roasting process and an intimate cafe experience. Finally, OpenTable isn't simply a tool to reserve tables; instead, it's a booking experience for diners to make online reservations and to help restaurant workers with real-time table management on OpenTable's provided computers and software. Carefully controlled, convenient experiences enabled these vertical businesses to succeed.

This is an intriguing phenomenon when we consider that after the PC revolution, a popular consensus among business thinkers was that vertical integration just doesn't work. The earliest example of vertical integration goes all the way back to Henry Ford.[24] The Ford company owned iron ore mines, made its own windshields, and had end-to-end control over everything from manufacturing to sales. For a while, the vertical model was wildly successful but eventually went out of fashion as suppliers became increasingly globalized and their prices became more competitive. Some would even argue that Ford "short-circuited" innovation. For example, Ford had no incentive to improve their windshields: they would all sell with Ford cars anyway. Meanwhile, to compete with Ford, third-party glass factories eventually had an incentive to create better windshields than Ford's.

After the days of Ford came and went and after Apple lost the PC war and nearly went bankrupt, business observers came to believe that vertical integration simply wasn't viable. Critics would blast the model because it effectively stalled innovation, like Ford did with windshields. Furthermore, vertical integration didn't seem as economically sustainable: why own and manage every core aspect of your operation when you can pay someone else who specializes in a trade to do things for you? However, the current global recession proved that horizontal integration also has vulnerabilities. During the financial crisis scores of cheap Asian manufacturers have shut down their operations, in turn hurting many horizontal companies as a result.

Now many companies are building their own skyscrapers.[25] Oracle, famous for a horizontal strategy of selling business software, is now striving to produce multiple aspects of the chain: chips, computers, data storage, and software. Pepsi just bought its own bottling company to have greater control over its beverage distribution. Downtrodden auto company General Motors is attempting to buy back Delphi, a supplier it spun off in 1999, as it emerges from bankruptcy. In 2009 Boeing bought a factory and a 50 percent stake in a joint venture to make parts for the troubled 787 Dreamliner Jet. Hewlett-Packard, a computer maker that usually relies on Microsoft to take care of its software needs, has acquired Palm to create an in-house OS for its smartphones and tablets. The list goes on. However, to say that all these companies are going absolutely vertical would be inaccurate. Similar to what Microsoft is doing with Windows Phone 7, these companies appear to be verticalizing the parts of their chain that truly matter to the experience they offer—in other words, parts that are likely to result in profits. Essentially these semivertical companies are leaning skyscrapers.

That the aforementioned companies aren't embracing absolute verticals is probably a good thing. Old concerns still remain about vertical integration: that a vertical company can have too much control over one niche, just like when owners of raw materials could squeeze competitors. So what happens when a company is too vertical today in the digital age, with digital resources?

This all brings us back to the question our discussion of OpenTable prompted. With its vertical business strategy, does OpenTable truly hold participating restaurateurs "hostage"? With the high costs of OpenTable's service and the lack of a sizable alternative, some restaurant owners certainly feel that way, and this element of concern highlights some new dangers of vertical integration in the digital age. Companies are now embracing the vertical power of digital resources, which are becoming increasingly valuable in today's always-on, information-driven economy.

Though Apple has delivered many benefits with its vertical strategy, the company poses threats as well. Apple is no longer simply a gadget maker and a music store. With the phenomenal success of the iPhone's App Store, Apple has become a powerful media publisher of software, books, magazines, newspapers, and more. What's more, Apple hasn't been shy about flexing its muscle or imposing rules on third-party programmers about what can and can't be inside their iPhone apps. Apple is no longer only seeking to control its own mobile destiny; rather, it's fighting for control over the destiny of digital media.

Chapter 5

the battle
for
control

The application had a simple premise: shake the iPhone to undress the girl. German tabloid *Bild* inserted this special gimmick into its iPhone app to allow users to undress *Bild*'s girl of the day in hopes of attracting more buyers. But Apple refused to allow the nude images into the app store, deeming sexual content objectionable.

Bild can't say they weren't warned. Apple owns and operates the App Store, so it is entitled to decide which apps are allowed inside it. The rule about nudity was clear from day one: when Steve Jobs introduced the App Store in June 2008, porn was at the top of the list of content that wouldn't be allowed inside, along with malicious content and apps that invade privacy. To *Bild*, however, Apple's rejection of its iPhone app was an act of censorship with troubling implications for press independence. In reaction, the publication wrote a letter urging the Federation of German Newspaper Publishers to fight Apple's non-erotic policy. "Today it is naked breasts," wrote Donata Hopfen, head of *Bild*'s digital media department, "tomorrow it could be editorial content."[1]

Many consider Apple's App Store to be the most controversial product Apple has ever made. Participating in the App Store comes with many limitations. In Jobs's effort to maintain control over the iOS platform, Apple has enforced stringent regulations around how exactly apps can be made, and App Store reviewers have also rejected a large number of apps they deemed inappropriate.

Though Jobs originally made it sound like any developer, big or small, had an equal opportunity to sell apps in the App Store, it is not that simple. In actuality, if a person makes an app for the iPhone, he has to make it Apple's way or it won't be offered in the App Store. He has to play by Apple's strict rules. Apple's role as a gatekeeper sets a precedent with disturbing implications for the future of creative freedom. "It's so hard to reconcile my love for these beautiful devices on my desk with my hatred for the ugly words in [the iPhone developers'] legal agreement," former Facebook employee Joe Hewitt said.

For those reasons, critics have accused the App Store of imposing censorship, holding developers hostage, stifling innovation, and fostering conformity. Apple unlocked a dream device for the masses, and it is doing everything within its power to keep it under control.

Apple must approve every iPhone app before it goes up for sale in the App Store, and this means that the corporation can regulate and censor content however it wishes. However, other than explicitly banning porn, viruses, and privacy-invading apps, Apple hasn't published rules about what types of content are allowed in the App Store. In the years I have been writing about Apple, I have seen all sorts of rejections: apps rejected for containing special phone functionalities, for satirically mocking famous people, and even for reasons that seem personal.

Vanity Fair writer Michael Wolff felt that Apple unfairly punished him when they rejected his iPhone app dedicated to displaying his personal columns. Because Wolff has been known to write critically about Apple and its famously mercurial CEO, to him, the rejection was personal. "I'm not exactly unknown to these people," Wolff told me on the phone. "When I did a piece on Jobs a couple of years ago in *Vanity Fair*, they were fairly ticked off. And I've ever since very clearly been pushed out as a journalist they don't want to deal with. That's what I figure is going on here."[2]

FreedomVoice Systems developed a call-forwarding app called Newber, which didn't get a response from Apple at all, staying in eternal limbo with the App Store's review team. After Apple ignored Newber for six months, FreedomVoice Systems tabled the project after spending $500,000 on developing and marketing the app. "We followed all guidelines set by Apple throughout the development process and have never received comment from Apple as to why the Newber application has still not even been reviewed," wrote Eric Thomas, CEO of Freedom-Voice Systems, in a letter to his staff. "Steve Jobs hailed the App Store as, 'the best deal going to distribute applications to mobile platforms.' Our experience is that it is the worst deal going."[3]

That's the trade-off of working with Apple. With the phenomenally popular iOS App Store, software creators have a higher chance of gaining exposure and making money than they do on competing app stores, but developers must follow Apple's technical and editorial rules in order to get into the store. In doing so, they relinquish control to a giant corporation. This is comparable to if Microsoft not only sold Windows but also owned every computer and every store in which it was sold as well as controlled every developer that wished to sell software for the computer.

Is this the end of creative freedom? Not quite. There is still the web, after all. Through a browser on any computing device, including the iPhone, you can access all the porn, satiric cartoons, and Apple-bashing articles you want. Furthermore, according to Jobs, only 5 percent of the fifteen thousand apps submitted each week are rejected.[4] However, *Bild*, Wolff, and other App Store critics have valid reasons to feel concerned. After all, Apple is the leader in the mobile app revolution: in 2009 Apple sold 99.5 percent of app downloads served in the entire mobile software industry. At this rate, Apple's conservative software requirements will inevitably have a strong influence on the future of media. We are only beginning to see the dangers of a single point of control. If Apple maintains its lead as a mobile platform, this could be detrimental to the future of editorial independence, creative freedom, and programming.

The Future of Media

"I'm from the media world, and as you may have heard, we have lots of questions about our future," *Wired* magazine chief Chris Anderson said during a presentation at TED 2010. "The good news is I think we found part of the answer."[5]

Allured by success stories of mobile app makers, revenue-shedding publishers, including Condé Nast, were eager to play ball with Apple's latest creation: the iPad tablet, the 9.7-inch touchscreen device succeeding the iPhone. *Wired* was an early adopter of the new medium, releasing an iPad edition of its magazine only two months after the device launched. "We think this is a game changer," Anderson told the TED audience.

Indeed, before Apple even officially introduced the iPad, many technology journalists, including me, preached about its potential to reinvent publishing. After all, what could be better for reviving dying newspapers and magazines than a shiny new gadget attached to iTunes, the most successful digital business platform of all time?

However, we did not predict that publishers' editorial content would be subject to the approval of Apple's temperamental App Store reviewers, like *Bild*'s was. What if *Wired* published content that Apple deemed "objectionable?"

We have yet to see a mainstream publication rejected or pulled for any content Apple deems objectionable; however, because of the logistics of embracing a new publishing medium such as the iPad, the issue is poised to grow as more devices sell. Media operations must integrate digital tablet production into their infrastructure, and obtaining the software developers, designers, and content creators to make such a transition is neither easy nor inexpensive. Furthermore, if advertisers invest more money in the iPad version of a publication, that in turn pressures publishers to give priority to resources allocated to the device. So, for example, if a *Time* magazine app sells better on the iPad than it does in print, then the company would likely produce content for the iPad first and then do so for print and the web. As a result, what we would potentially get as readers of all media would be the censored, "App Store safe" version of the content.

Nevertheless, iPad apps certainly might become the revenue goldmine that publishers are seeking: right after the iPad launched, a number of large corporations paid between $75,000 to $300,000 to place ads in iPad apps for *Newsweek*, Reuters, *Wall Street Journal*, and other major publications. To further illustrate that digitized publications will be serious business, the Magazine Publishers of America conducted research that found that 60 percent of US customers are expected to purchase an e-reader or a tablet in the next three years.[6] Given Apple's lead in mobile technology, the rate at which Apple and the App Store are growing, and the wild enthusiasm among advertisers lining up for the iPad opportunity, inevitably Apple will, to some extent, have influence over the content that publishers produce.

This story will sound familiar to anyone who has followed Walmart and the music industry. Walmart refuses to sell music albums carrying the Parental Advisory tag, and in the past the megaretailer has occasionally suggested that artists change lyrics and CD covers it deems objectionable. Given the retail chain's position as the world's largest brick-and-mortar music retailer, many agree that Walmart has altered the way the recording industry creates albums. In the production process many musicians and record companies decide whether to sanitize lyrics and album covers in order to gain Walmart's approval. To avoid conflicts, big record labels will often issue two versions—one "sanitized" version for Walmart and another unedited—but labels do this only for their star artists. Smaller artists who don't get two versions of their CD feel the most pressure: either scrub yourself for Walmart or sacrifice an opportunity to reach the masses.

In the case of Apple, in addition to putting our content at the mercy of a team of reviewers, we also must account for Apple's technical requirements for apps, which are subject to change whenever the company wishes. In fact, Anderson's iPad app was nearly a victim of Apple's fickleness. *Wired* has a long-standing partnership with Adobe, who provides the software (InDesign) and custom plug-ins used to design its beautiful magazine pages, and the two parties entered an agreement to produce a tablet edition of the publication together with Adobe Flash. However, only two months after Anderson's TED talk introducing the *Wired* tablet app, Apple revised its iOS programmer agreement, prohibiting usage of third-party programming languages for creating apps. That effectively blocked Adobe's Flash code, which temporarily threw a wrench into *Wired*'s plans. Ultimately, though, *Wired*'s iPad app made its way into the App Store after the magazine pushed Adobe to repurpose the app with Apple's approved programming language, Objective-C.[7]

Apple's rule prohibiting usage of third-party languages sparked a firestorm of controversy among creators. The outcry even provoked Steve Jobs to pen an open letter famously titled "Thoughts on Flash," in which the CEO said Apple was blocking Flash to protect mobile innovation.[8] He explained that Flash programs were made for computers with mice and keyboards, not touchscreen devices like the iPhone, and that its performance was shoddy, causing frequent crashes and rapid battery drainage on the Mac. "We know from painful experience that letting a third party layer of software come between the platform and the developer ultimately results in sub-standard apps and hinders the enhancement and progress of the platform," Jobs wrote. "If developers grow dependent on third party development libraries and tools, they can only take advantage of platform enhancements if and when the third party chooses to adopt the new features. We cannot be at the mercy of a third party deciding if and when they will make our enhancements available to our developers."

Later in 2010, for reasons unknown, Apple undid its prohibition on apps coded in third-party languages. However, iPhone programmer Hampton Catlin told me he had received a phone call from the FTC, which was investigating claims of anticompetition in light of Apple's ban on third-party programming tools.[9] This suggests that Apple likely relaxed its restrictions after raising flags with the FTC.

But then in February 2011, Apple issued a new rule, this time explicitly and specifically for publishers. Apple created a new system for publishers to sell subscriptions to their magazines, newspapers, video, or music through apps, and the rule was such: Apple keeps 30 percent of each subscription sale sold through an app. So, for example, if a customer purchased a subscription of the iPad edition of *Wired* magazine, the publisher keeps 70 percent of the subscription sale, and the rest goes to Apple. The 70–30 split is the same that Apple shares with any app programmer selling apps in the App Store, but the subscriptions policy later gets more confusing and vague.

The subscriptions policy states that publishers selling subscriptions must use Apple's proprietary in-app purchase system to sell them. In the past some publishers were directing iPhone, iPad, or iPod Touch customers to external web links to purchase individual books or magazines without going through the App Store. Apple's new rule requires publishers to sell directly within the app; publishers can continue to sell through their web store, but a URL to the web store is not allowed inside the app. Furthermore, outside the App Store, a publisher may not undercut the price of the subscription offered inside the app. Finally, publishers were not allowed to collect user data from in-app subscribers without a user's approval.

"Our philosophy is simple—when Apple brings a new subscriber to the app, Apple earns a 30 percent share; when the publisher brings an existing or new subscriber to the app, the publisher keeps 100 percent and Apple earns nothing," said Steve Jobs, Apple's CEO, in a press statement. "All we require is that, if a publisher is making a subscription offer outside of the app, the same (or better) offer be made inside the app, so that customers can easily subscribe with one-click right in the app. We believe that this innovative subscription service will provide publishers with a brand new opportunity to expand digital access to their content onto the iPad, iPod touch and iPhone, delighting both new and existing subscribers."[10]

The benefits of the subscriptions policy for customers are obvious: they get a faster and more seamless experience when purchasing a subscription from directly inside an app. But from a publisher's perspective, the policy seems unfair. There are plenty of software companies that provide monthly services through iPhone, iPod Touch, or iPad apps. However, software service providers are not required to use Apple's in-app payment system or give up 30 percent of subscriptions, and publishers are. Why not? What's the difference between a publisher and a software service provider if both are using an app channel to sell their product? Apple is treating publishers differently from regular app programmers in this regard, which makes the App Store an unfair playing field. Apple will likely change the subscriptions policy if publishers aren't pleased, but the initial policy requires publishers to give up control over pricing as well as direct access to their customers. "With its enviable tandem of products that everyone wants and a store where everyone buys, Apple believes it has the leverage to tell publishers: They are our customers. You can rent them, but they will remain in our custody," said David Carr of *New York Times*.[11]

Some of Apple's other policy rules are just as stringent. The Electronic Frontier Foundation (EFF) acquired and published the once-confidential iOS developer agreement using the Freedom of Information Act.[12] The contract, which all developers are required to sign in order to serve apps through the App Store, is surprisingly one-sided:

- Apple bans public statements, forbidding developers to speak about the agreement (which is why the Freedom of Information Act was needed to obtain it).
- Apps made with the iPhone software development kit can only be distributed through the official App Store, not unauthorized app stores opened by hackers.
- Apple's liability to a developer is limited to $50.
- If a third party sues Apple because of the developer's actions, then Apple could seek to recoup all amounts from the developer.
- Reverse engineering, or enabling others to reverse-engineer, the iPhone SDK is not allowed.
- Modifying Apple products is not allowed. This also means that apps that enable modifying or hacking Apple products are not allowed.
- Apple can "revoke digital certification of any of Your Applications at any time," which means an app can be pulled even if it has already been approved.

In short, developers' iOS creations basically belong to Apple because the developer can't offer them anywhere else. "If Apple's mobile devices are the future of computing, you can expect that future to be one with more limits on innovation and competition than the PC era that came before," said Fred von

Lohmann, a senior staff attorney with EFF. "It's frustrating to see Apple, the original pioneer in generative computing, putting shackles on the market it (for now) leads."[13]

In particular, the rules guarding how applications are programmed are alarming. A strong appeal for Apple's iPad is that because it is so easy to use, it could very well be the first computer parents buy for their children. However, the iPad also is so restricted that many current programmers believe that it could mean the end of hacking.

Tinker Me Not

In 1968, when computers still weighed over a hundred pounds and ate punch cards, Xerox PARC researcher Alan Kay began dreaming up the perfect portable computer. He fashioned models out of cardboard, lining the cutouts with lead pellets to simulate different sizes and weights. Eventually Kay crafted a very thin, tablet-like device with a large screen at the top and a keyboard at the bottom. He determined that this computer would have to weigh no more than two pounds in order for it to be comfortably portable, and most importantly, it would need to be highly dynamic and easy to use so that children of all ages could learn programming and science. Kay dubbed this concept the Dynabook.[14]

The Dynabook was never made, but historians agree that Kay's concept strongly inspired the mobile devices we tote around today. Fifty years after Kay published his concept, Apple's iPad—a 1.5-pound tablet computer with a 9.7-inch touchscreen display and an extremely simple interface—comes closest to matching Kay's description of the Dynabook. But Kay told me he doesn't believe the iPad is quite there. It may deliver on the tablet part of the Dynabook, but it does not meet the educational aspect because of Apple's strict rules on programming for the iPad. Apple's gatekeeping policies affected even Kay, a personal friend of Steve Jobs. In mid-2010 Apple blocked the Scratch app, a kid-friendly programming language based on Kay's work, from getting onto the App Store. The Scratch app displayed stories, games, and animations children made using MIT's Scratch platform, which was built on top of Kay's programming language Squeak. The news did not please Kay. "Both children and the Internet are bigger than Apple, and things that are good for [the] children of the world need to be able to run everywhere," Kay told me.[15]

Why are kid programmers so important to Kay? These up-and-coming digital creators are the ones who will be writing the technology that provides our everyday services and content, especially in a culture that's more connected than ever before. To put if very broadly: if Apple discourages these kids from learning on the most innovative technology in the market, it is stifling future innovation.

John McIntosh, a software developer unaffiliated with MIT who authored the Scratch app, said that Apple removed the app because it allegedly violated a rule in the iOS developer agreement, clause 3.3.2, which states that iOS apps may not contain code interpreters other than Apple's.[16] The clause reads,

> An Application may not itself install or launch other executable code by any means, including without limitation through the use of a plug-in architecture, calling other frameworks, other APIs or otherwise. No interpreted code may be downloaded or used in an Application except for code that is interpreted and run by Apple's Documented APIs and built-in interpreter(s).[17]

"If you follow the chain of where Scratch came from, yes it is a Dynabook app, sadly not an iPad app," McIntosh said. He and some members of MIT's Scratch community wrote letters asking Apple to reinstate the app. However, even if Apple does welcome Scratch back into the platform, the closed, controlling nature of the App Store still sets a negative precedent for the future of programming, critics say. "I think it's terrible," said Andrés Monroy-Hernández, a PhD candidate at the MIT Media Lab and lead developer of the Scratch online community. "Even if the Scratch app was approved, I still think this sends a really bad message for young creators in general. We have a forum where kids post comments, and they were really upset about this."[18]

Some critics contend that Apple's imposed control over how apps can be programmed foreshadows the end of tinkering. Software programmer and self-proclaimed hacker Mark Pilgrim reminisced about the days when personal computers were truly "personal," meaning a user could do anything he wanted with his device without feeling like a rule-breaker. "You could turn on the computer and press Ctrl-Reset and you'd get a prompt. And at this prompt, you could type in an entire program, and then type RUN, and it would motherfucking run," he said.[19]

Twitter engineer Alex Payne echoed similar concerns. "The thing that bothers me most about the iPad is this: if I had an iPad rather than a real computer as a kid, I'd never be a programmer today," he said. "I'd never have had the ability to run whatever stupid, potentially harmful, hugely educational programs I could download or write. I wouldn't have been able to fire up ResEdit and edit out the Mac startup sound so I could tinker on the computer at all hours without waking my parents. . . . Perhaps the iPad signals an end to the 'hacker era' of digital history."[20]

But maybe not. With each new version of the iOS operating system, a small army of independent programmers and hackers get to work prying it open, removing restrictions to make their devices run unauthorized software and work with different carriers. Doing so has become somewhat of a cat-and-mouse game: the jailbreaker hacks an iOS product, Apple issues a software update to disable these hacks, and then jailbreakers find another way in. In fact, the first person to ever unlock the iPhone to run on different carriers was George Hotz—when he was seventeen years old.[21] Clearly, the hacker era is far from over.

What is different today, however, is the sentiment surrounding hacking. In 2009, in a request filed to the US Copyright office, Apple tried to make hacking the iPhone illegal, claiming that hacking the device opens doors to viruses, invasion of privacy, and pirated software. A year later, however, the copyright office decided that jailbreaking was legal.[22] Still, Apple refuses to support jailbroken iPhones—if you hack the device, you void your warranty. So in the very least Apple has made hacking impractical, and the developer rules suggest that developers who share their apps with unauthorized app stores can get their licenses revoked. Ultimately, Apple creates the sentiment that hacking is a criminal activity, which may deprive new programmers of the incentive to explore, experiment, and, in turn, innovate. This development is alarming for the future of the web and digital software in general because, as Payne and Pilgrim point out, the world's best programmers are hackers as well. "The iPad may be a boon to traditional education, insofar as it allows for multimedia textbooks and such, but in its current form, it's a detriment to the sort of hacker culture that has propelled the digital economy," Payne said.

Overall, Apple's control-freak lockdown on its iOS platform was bound to attract criticism, and even some of Apple's partners started to turn cross. Google couldn't resist delivering some of the first pot shots.

An "Open" Alternative?

The friendship seemed so good at first. When Steve Jobs introduced the iPhone in 2007, he proudly boasted that the handset would feature Google's maps and search services. With a big smile on his face, he welcomed his friend Eric Schmidt, Google's then CEO, to come on stage.

"Congratulations, Steve! What an incredible job," Schmidt said in front of an audience at Macworld 2007. "So Steve, I've had the privilege of joining [Apple's] board, and there's a lot of relationships between the boards. And I thought if we could just sort of merge the companies, we could call them Applegoo, but I'm not a marketing guy."[23]

A year and a half later, that Schmidt wasn't remotely interested in merging with Apple became clear when the first Google-powered phone hit the market. In 2006 Google had acquired Android, a mobile start-up based in Palo Alto, but at the time Google's plans were vague. The Android mobile operating system debuted on an HTC-manufactured device called the G1, and like the iPhone, it was a general-purpose device that sported a touchscreen with tappable icons and supported third-party apps. Members of the media were quick to label the Android G1 an "iPhone killer." Furthering its competition with the iPhone, Google later released the Nexus One, the first Google-branded phone, which the search giant designed with HTC. Not surprisingly, Schmidt later resigned from Apple's board of directors because of "conflicts of interest."[24] Incensed by his competitor, Jobs spoke to his employees at an off-the-record town hall meeting as if he were a military leader rallying troops to go to war.

Over the last few years the two giants have found themselves butting heads again and again. These weren't simply two rivals competing in the mobile industry; these were two giants fighting over control of the destiny of media. To date, both companies have started units, formed partnerships, or made acquisitions in several of the same arenas: digital books, web browsers, PC operating systems, music, advertising, maps, search, television, personal broadcasting, and mobile advertising. The two bid on some of the same companies as if to spite one another—for instance, AdMob, a mobile advertising company that sold to Google, and Palm, pioneer of the PDA, which sold to HP. "We did not enter the search business," Jobs said. "They entered the phone business. Make no mistake they want to kill the iPhone. We won't let them." An employee attempted to change the topic, but there was no getting Jobs off this rant. "I want to go back to that other question first and say one more thing. This 'don't be evil' mantra: It's a load of crap."[25]

Google, however, wasn't the type to turn the other cheek, and the company was bold enough to trash talk its rival on stage at its 2010 Android developers conference. "If Google did not act, we faced a Draconian future where one man, one company, one device, one carrier would be our only choice," Google vice president of engineering Vic Gundotra told a crowd as he stood in front of a projected slide depicting George Orwell's 1984. "If you believe in openness, if you believe in choice, if you believe in innovation from everyone, then welcome to Android."[26]

Google's main weapon against Apple was its original commitment to openness and choice. Virtually all of Google's Internet services would be available to any type of device equipped with a web browser, whereas most of Apple's media services require Apple-made hardware. The Android OS was open source, meaning programmers can freely hack the source code to unlock special functions such as Internet tethering. For mobile apps Google did host its own app store, Android Market, which, like Apple, also prohibited porn, malware, and apps that invade privacy, but unlike the App Store, the Android Market allowed users to "sideload" any unauthorized apps that developers distributed in unofficial stores or portals online. Last but not least, Google launched the Open Handset Alliance, an allegiance with hardware makers (not carriers) to make phones that were open, elegant, powerful, and not subject to the legendary whimsies of wireless carriers, which are known to cripple devices, fail to innovate on new features, charge extra for whatever built-in features they do include, and control what can and can't be done with the devices. Furthering its commitment to "openness" on the web, Google also released a lightweight PC operating system called Chrome, which is a modified browser that runs web apps and can work with any PC loaded with Google's Chrome web browser. The list goes on, and considering the number of clashing differences between Apple and Google, in hindsight, that Schmidt was ever on Apple's board of directors to begin with is shocking.

The good news is that Google provides more open alternatives for programmers and media creators to distribute their content. The bad news is that Google's open modus operandi in the wireless world is steadfastly failing, and the search giant has already surrendered control to the carriers.

Google's track record for maintaining its seemingly philanthropic "open" regime was flawless until August 2010, when many observers lost faith in the search giant's "Do No Evil" mantra. At that time, in collaboration with Verizon, Google drafted a proposal that would effectively subject the mobile Internet to corporate control.

The Internet and the "Internot"

For years, in a movement popularly known as "net neutrality," Internet activists have pushed for laws to keep the Internet an open playing field. The principle proposes that neither Internet service providers nor the government can set restrictions on any type of content, service, or method of communication on the Internet or on any of the equipment attached. Proponents of net neutrality fear a future in which corporations and the broadband industry can block certain types of Internet applications or content to hurt competitors. Among many supporters, Google was the biggest corporate advocate of net neutrality—until August 9, 2010.

Google was for years the biggest corporate advocate of net neutrality. Google and ISPs were natural enemies; for Google, the math was simple: the cheaper and faster Internet service is, the more people will use the net, which in turn leads to more ad profits for Google. For ISPs, they had spent years building their networks, only to see the real profits go to companies like Google, whose services run through the ISP's pipes to subscribers. The real money, the ISPs were finding, isn't in being a utility but rather in offering extra services to subscribers.

But Google's relationship with ISPs and mobile carriers began to change with the Android smartphone operating system, where the search ad-giant needed the carriers to adopt Android devices so, in turn, Google allowed carriers to bundle their special services in the Android-powered smartphones they sell.

That new relationship led to an unexpected proposal on August 9, 2010. In a document titled "A proposal for an open internet," Google and Verizon outlined a framework that they claimed would "protect the future of openness" by ensuring that no Internet traffic of any kind is prioritized over any other kind.[27] The wording of the seven-tier proposal was ambiguous.

They proposed a framework that requires allowing customers to

1. send and receive lawful content of their choice;
2. run lawful applications and use lawful services of their choice; and
3. connect their choice of legal devices that do not harm the network or service, facilitate theft of service, or harm other users of the service.

The term "lawful" would open the door for Internet service providers to engage in antipiracy snooping, just as the recording and movie industries have been pushing for years, even though deciphering via computer what's pirated content and what's legitimate is nearly impossible. Furthermore, rather than the FCC, Google and Verizon suggest that "independent" third parties would enforce rules against those who are being unlawful. In response, Karl Bode, editor of DSLReports.com, pointed out that these third parties would likely be groups the telecom industry itself would create: "faux-regulatory agencies created by AT&T, Verizon and their massive lobbyist coalitions of hijacked political groups, paid policy mouthpieces and fake consumer advocates."[28]

In the agreement Google and Verizon state,

The FCC would enforce the consumer protection and nondiscrimination requirements through case-by-case adjudication, but would have no rule making authority with respect to those provisions. Parties would be encouraged to use non-governmental dispute resolution processes established by independent, widely recognized Internet community governance initiatives, and the FCC would be directed to give appropriate deference to decisions or advisory opinions of such groups.

Furthermore, Google and Verizon suggest segmenting a separate channel for "additional, differentiated online services" for broadband providers to develop new services, such as health care monitoring—in other words, a private segment of the Internet, which contradicts their stated purpose of avoiding prioritization of any kind of content on the Internet.

Finally, in one extremely vague clause of the proposal, the companies recommended treating the wireless mobile Internet as a separate channel in which these rules of openness do not apply. "We both recognize that wireless broadband is different from the traditional wireline world, in part because the mobile marketplace is more competitive and changing rapidly," wrote Alan Davidson, Google director of public policy, and Tom Tauke, Verizon executive vice president of public affairs, policy, and communications. "In recognition of the still-nascent nature of the wireless broadband marketplace, under this proposal we would not now apply most of the wireline principles to wireless, except for the transparency requirement."

Despite Google's strong track record of adhering to its open regime, net neutrality proponents point out that the proposal is packed with loopholes that would effectively end the open web by dividing it into two parts.

The most disconcerting tidbit from the proposal is the vague clause about the "wireless" web. Why would Google and Verizon want zero rules of "openness" to apply to Internet for phones, or why should there be separate Internet for mobile devices at all? Whatever their intentions may be, the wording of the proposal would effectively enable Google and the telecom industry to regulate Internet connections for phones to their own benefit. For instance, Google could strike deals to have YouTube video stream faster than their competitors' videos, or telecom companies could speed up services that they offer while slowing down or raising costs for smaller companies wishing to offer services for web-connected phones.

In short, Google, Verizon, and other big players in the telecom industry could potentially reserve for themselves high-speed, expensive chunks of the mobile web. This would render smaller, less wealthy startups or independent programmers making apps for phones competitively crippled by a much slower connection to deliver their content and high costs to provide these services. Is this a blow to innovation?

We will find out over the next few years. A few months after Verizon and Google published their proposal, the FCC passed net neutrality rules that strengthen regulations against wired broadband but do not formally extend all neutrality protections to wireless networks. The FCC did go further than Google and Verizon suggested, forbidding mobile carriers from outright blocking websites; requiring that customers be able to use, via their smartphones, Internet-based calling services such as Skype; and prohibiting companies like Verizon from playing favorites with online video sites. Nonetheless, the FCC's rules have many exceptions and lack the strength of the rules that apply to DSL and cable companies, leaving many thinking that the door is wide open for wireless networks to create high-priced fast lanes for premium customers, and this could give these networks a chokehold in the future.

"The agreement between Verizon and Google about how to manage Internet traffic is nothing more than a private agreement between two corporate behemoths, and should not be a template or basis for either Congressional or FCC action," said Gigi B. Sohn, president and cofounder of Public Knowledge, a public interest group fighting for citizens' rights in the emerging digital age. "It is unenforceable, and does almost nothing to preserve an open Internet. Most critically, it sacrifices the future of the mobile wireless Internet as this platform becomes more central to the lives of all Americans."[29]

So here we have two giants contending for the destiny of media: Apple, aiming to own exclusively a set of media channels to create an experience that sells hardware, and Google, a company that has positioned itself to have an advantage on a regulated version of the mobile web. With the App Store, Apple regulates the types of content third parties produce and how they produce it, and although Google may offer a more open alternative in that regard, the search giant's proposal with Verizon shows an intention to create a separate, closed chunk of the Internet, thereby putting it in position to dominate against smaller content creators. As much as Google toots its "openness" horn, it is also advocating a closed segment of the web.

The Soul of Mobile

An important question to ask is, Why does Apple have to be so controlling? Why all the stringent rules about how apps are made and which apps are allowed to run on the iPhone, iPad, or

iPod Touch? That the key move from the start was wrestling control away from the carriers is important to remember. In the past, carriers controlled almost everything related to mobile experiences: they determined which software features would ship with phones and how much it would cost for extra ones to be enabled. Evidently, carriers want some of that control back.

Elia Freedman, CEO of Infinity Softworks, who has worked in the mobile software business for thirteen years, says there is a serious fight going on over the soul of mobile. "We have been distracted by thinking that Google is Microsoft and Apple is Apple in a doomed fight already fought 20 years ago," Freedman wrote. "But that is not the fight we should be caring about at all. The fight we should be talking about, but aren't, is the fight between mobile device makers and the carriers. This is the only real fight that matters."[30]

After initially ceding control to Apple in order to create the iPhone experience, carriers are fighting back. Google Android phones illustrate this problem. Some Android phones ship with bloatware—software that the carrier forces onto the handset that customers can't remove. And then there is the Nexus One, Google's flagship smartphone that was sold unlocked to work on any carrier; the device even shipped with an unlockable bootloader that allowed programmers to go directly into the open source of the smartphone in order to hack it to do whatever they want. However, after carriers refused to service the Nexus One, Google was forced to discontinue the device because of poor sales.

So even though critics have described Apple's tight control over the iPhone platform as "draconian," it does have some benefit. As customers, we had less freedom before when carriers were at the steering wheel. Although app developers sacrifice some control to companies like Apple and Google, at the end of the day, having an opportunity to sell mobile apps grants them more freedom than they had before 2007.

"This is war," Freedman wrote. "And this war will go nothing like Apple versus Microsoft. This is about who controls the experience; who gets to interact with the customer." Looking to the near future, perhaps the question we should ask is whether media on our always-connected gadgets imprisons us as individuals. What are the implications of having anything, anytime, and anywhere on our brains? Are we getting smarter or dumber? More functional or hopelessly dependent? Happier or more depressed? The answers are hardly simple.

Chapter 6

smarter or
dumber?

When he turned nineteen, Norwegian Magnus Carlsen became the youngest chess player in history to be ranked number one in the world. By age thirteen he had already become a grandmaster.[1] Despite these accomplishments, he doesn't consider himself to be a genius in the traditional sense; in fact, in press interviews Carlsen says he is afraid that knowing too much could be a curse. To illustrate his point, Carlsen cites John Nunn, one of England's strongest chess players who never became number one. "I am convinced that the reason the Englishman John Nunn never became world champion is that he was too clever for that," Carlsen explained. "At the age of 15, Nunn started studying mathematics in Oxford; he was the youngest student in the last 500 years, and at 23 he did a PhD in algebraic topology. He has so incredibly much in his head. Simply too much. His enormous powers of understanding and his constant thirst for knowledge distracted him from chess."[2]

In contrast, Carlsen describes himself as lethargic and sloppy. "Organization is not my thing. I am chaotic and tend to be lazy," he said. When he feels good, he trains a lot. When he feels bad, he doesn't bother.

This chessmaster doesn't sound very different from the average teenager. In fact, like most teens, Carlsen spent much of his free time playing with a computer in between other time wasters such as assembling jigsaw puzzles or watching TV. When he was eleven, he began preparing for chess tournaments by playing against other players online.

Carlsen credits his achievement as world champion to his Internet experience. He claims that he learned faster than previous generations—an edge he gained thanks not only to the Internet but also his family. When Carlsen was thirteen, his father yanked him and his sisters out of school for a year. Carlsen's parents schooled him and his sisters while traveling by car to Austria, Montenegro, Greece, Italy, and Hungary. "That was fantastic, much more effective than sitting in school," Carlsen said of his year traveling when he was thirteen. "I do understand that it is a problem for a teacher having to look after 30 pupils. But the slow speed was quite frustrating for me. I didn't miss school at all."

This young chessmaster serves as a strong case study for the web-affected youth growing up in a culture that is always on. He didn't become a chess prodigy because of systematic learning in a classroom; rather, the chaos, disorder, and inconsistency throughout the digital world nurtured him. By playing chess online, Carlsen exposed himself to strategies used by all sorts of different players from all over the world of various ages, ethnicities, and skill sets. For this reason, he predicts that with this same edge of being able to learn faster, even younger children will dethrone him soon. "Nowadays, children start using a computer at an even earlier age; they are already learning the rules on screen. In that sense I am already old-fashioned. Technological progress leads to younger and younger top players, everywhere in the world."

However, despite triumphant stories like Carlsen's, a growing chorus of voices agrees that always-on technology is destroying our ability to concentrate and, as a result, making us dumb. Critics have cast web-connected laptops, smartphones, and video game consoles as distracting brain mushers. Even famously tech-savvy President Barack Obama is voting against this very phenomenon. "You're coming of age in a 24/7 media environment that bombards us with all kinds of content and exposes us to all kinds of arguments, some of which don't always rank that high on the truth meter," he told students during his visit to Hampton University. "And with iPods and iPads, and Xboxes and PlaySta-tions—none of which I know how to work—information be-comes a distraction, a diversion, a form of entertainment, rather than a tool of empowerment, rather than the means of emancipa-tion. So all of this is not only putting pressure on you; it's putting new pressure on our country and on our democracy."[3]

That Obama of all people would make such a claim about the destructive potential of the always-on digital age is, perhaps, ironic. In his campaign's 24/7 digital blitz of web videos, a social networking website, a Twitter account, and even an iPhone app, Obama vaulted over opponent John McCain to become the first black president of the United States. He accrued $600 million in funding from more than three million people, many of whom do-nated through the web.[4] And it didn't stop there. His online supporters posted thirty thousand events promoting him, and an iPhone app that developer Raven Zachary created also pushed people for votes.[5] The Internet empowered Obama's campaign; he used social media more effectively than any previous national candidate to win the election. Yet here, just two years into his presidency, Obama cautions against the phenomenon that got him where he is today.

Because carrying an Internet-connected portable computer everywhere means that we can gorge on just about every kind of media, regardless of where we are or what time it is, critics assume that being always on culminates in massive information overload. In this regard, the daily human task that researchers have been studying is multitasking—namely, the act of rapidly switching between tasks while incoming bursts of data such as e-mails, text messages, and IMs bombard us. As overwhelming as the always-on lifestyle sounds, a closer examination of contemporary psychological studies shows that there is no consistent data indicating negative effects on the brain.

Fractured Concentration?

Chicago resident Matt Sallee's life is a never-ending sprint that mostly takes place in his phone. At five in the morning the alarm goes off, and during his train commute the twenty-nine-year-old rolls through fifty e-mails he received overnight on his Black-Berry.

As a manager of global business development at an LED company, Sallee works in time zones spanning across three continents. Throughout the morning and day he shoots off e-mails and text messages to his boss in Italy and customers in Japan, building up toward an evening of conference calls with associates in Taiwan. Sallee often flies around for work too, keeping a camera with him to snap pictures of his travels that he sells on his website as an independent side business.

On top of all that, Sallee is a husband and a dog owner. His wife is also an extremely busy person, which is part of what makes their relationship work. However, when they are alone together, they are alone together; no phones or talking about work are allowed during dinner or their weekend dates spent hiking, biking, or snowboarding. "I love having 10 different things cooking at once, but for me it's all moving in little pieces, and when it comes time that there are big deliverables needed, I don't have to scramble at the last minute," Sallee said. "It's an hour of combining all the little pieces into one thing, and it's done."[6] Sallee likes to think of himself as a fantastic multitasker despite numerous reports suggesting that carrying a smartphone everywhere is destroying people's abilities to concentrate, fracturing social ties, and changing the way our brains are wired.

That the "always-on network" is eradicating the borders between home and office, thereby changing the way people work and play, is not news. But how much distraction can one person take? Research is still in the early stages, but there is little hard evidence that 24/7 access to information is bad for you. Nonetheless, the image of frantic, distracted workers scrabbling harder than ever for ever-diminishing social and economic returns is an attractive target for critics.[7]

Although seeing people chatting on cell phones in the popcorn line at the cinema is annoying, are these devices—and the multitasking they encourage—taking a massive toll on our psyches and perhaps even fundamentally altering the way our brains are wired? Is the smartphone—like Google, TV, comics, and the movies before it—actually making us dumb?

Stanford professor Clifford Nass made an intriguing discovery in 2009.[8] Working with a group of college undergraduates, Nass set out to study the effects of media multitasking on concentration and task switching. First, he administered a questionnaire to 262 students, asking them to self-report the amount of time they spent on a dozen media activities, including reading books, watching television, viewing online videos, playing videogames, listening to music, and more. Nass identified subjects who met a certain threshold as heavy multitaskers; those below were light multitaskers. After dividing the groups, Nass proceeded with several tests.

One test asked students to recall briefly glimpsed orientations of red rectangles surrounded by blue rectangles; the students had to determine whether the red rectangles had shifted in position between different pictures. The heavy multitaskers struggled to keep track of the red rectangles because they were having trouble ignoring the blue ones.

To measure task-switching ability, another test presented participants with a letter and number combination, like "b6" or "f9," and were asked to do one of two tasks: first, participants were instructed to hit the left button if they saw an odd number and the right for an even; then, they were asked to switch tasks, this time pressing the left for a vowel and the right for a consonant. They were warned before each letter-number combination appeared what the task was to be, but regardless of this alert, heavy multitaskers responded on average half a second more slowly than light multitaskers when the task was switched. In summary, the Stanford study found that people who were rated heavy multitaskers were less able to concentrate on a single task and also worse at switching between tasks than were light multitaskers. "We have evidence that high multitaskers are worse at managing their short-term memory and worse at switching tasks," Nass told me.

The Stanford media-tasking study was the most discussed social science story of 2009, and naturally, plenty of skeptics have scrutinized the results. For some, there remains the open question of causation: is multitasking hurting people's abilities to concentrate and switch tasks, or were these subjects already distractable people who were predisposed to become heavy multitaskers? The study's results were also open to interpretation: in the rectangles experiment unnecessary elements could easily distract these students, or the students might have a wider net of attention and are more able to pick up peripheral information. Last but not least, the sample size was rather small: after drilling through the 262 students who responded to the questionnaire, the study invited only 15 light multitaskers and 19 high multitaskers for the experiments. "Given that the study aimed to understand the effects of interactions among 12 different forms of 'media,' it's troubling that there were only 19 'High Media Multitaskers' tested, and that we don't know what sorts of media multitasking they actually reported engaging in," said Mark Liberman, editor of *Language Log,* a University of Pennsylvania linguistics blog.[9] In Internet forums Nass even acknowledges concerns about the small sample size and says follow-up studies are being conducted to further examine multitasking effects on the brain.

Regardless that research on media multitasking is still largely uncharted territory, some media writers are eager to draw a conclusion. Nicholas Carr cites Stanford's multitasking study as a key element in his book *The Shallows: What the Internet Is Doing to Our Brains.*[10] Using Nass's findings, Carr asserts that the Internet is shattering our focus and restructuring our brains to make us shallow thinkers. Because of this constant flow of data flowing in and out of our everyday lives, Carr argues that less information is traveling from working memory to long-term memory, and this dilutes our thinking and our ability to learn.

Following the release of Carr's book, the *New York Times* ran a feature series titled *Your Brain on Computers,* a collection of stories examining the negative consequences of gadget overload. Examples of headlines included "Digital Devices Deprive Brain of Needed Downtime,"[11] "Attached to Technology and Paying the Price,"[12] and "More Americans Sense Downside to Being Plugged In."[13] The author of the series, Matt Richtel, draws on research studies unrelated to human multitasking, such as a study showing that rats need to rest in order to absorb new experiences, in order to support his thesis that multitasking with always-on gadgets is damaging the human brain.

What's at stake here is a set of major choices about social policy and personal lifestyle. If it's really true that modern digital multitasking causes significant cognitive disability and even brain damage, as Richtel claims, then many very serious social and individual changes are urgently needed. Before starting down this path, we need better evidence that there's a real connection between cognitive disability and media multitasking.[14]

Research on media multitasking is still at an early stage, but one study is already challenging the notion that human brains are not meant to multitask. A University of Utah study published in 2010 suggests that some human beings are excellent multitaskers, whom the study labeled "supertaskers."[15] Researchers Jason Watson and David Strayer put two hundred college undergrads through a driving simulator, in which they were required to "drive" behind a virtual car and brake whenever its brake lights shone while at the same time performing various tasks such as memorizing and recalling items in the correct order and solving math problems.

Watson and Strayer analyzed the students based on their speed and accuracy in completing the tasks, and the researchers discovered that a small minority—just 2.5 percent (three men and two women) of the subjects—showed absolutely no performance decrements when performing single tasks versus dual tasks. Rather, these few individuals excelled at multitasking. Additionally, contrasting with the results of Stanford's study, the supertaskers were better at task switching and performing individual tasks than the rest of the group.

The rest of the group, however, did show overall degraded performance when handling dual tasks compared to a single task, suggesting that the vast majority of people may indeed be slower when processing multiple activities. But the discovery of supertaskers argues against the popular theory that the human brain is not meant to multitask, Watson and Strayer say. "Our results suggest that there are supertaskers in our midst—rare but intriguing individuals with extraordinary multitasking ability," Watson and Strayer wrote. "These individual differences are important because they challenge current theory that postulates immutable bottlenecks in dual-task performance."

Keeping in mind that the 24/7 multitasking lifestyle isn't new at all is worthwhile. Even though most of us probably aren't "supertaskers," we've been multitasking our entire lives in the everyday world. We can dribble a basketball while running, jot down notes while listening to a lecture, and jog through the park while listening to music. In fact, Vaughan Bell, a clinical and neuropsychologist at the Universidad de Antioquia, Colombia, argued that the most common, information-dense, and attention-demanding task is caring for children. "If you think Twitter is an attention magnet, try living with an infant," Bell said. "Kids are the most distracting thing there is, and when you have three or even four in the house it is both impossible to focus on one thing, and stressful, because the consequences of not keeping an eye on your kids can be frightening even to think about."[16] Kids are indeed distracting: a British study found that for drivers, the distraction of squabbling kids can slow down brake reaction times by 13 percent—as much as alcohol.[17]

Bell added that people in poorer neighborhoods that use very little technology (like Medellín, Colombia, where he resides) hardly live distraction-free lives. They have to watch their food because there is no timer; washing clothes has to be done by hand while keeping an eye on everything else; and when a street vendor passes by the house and shouts what they are selling, a family could go without food for a day if a person misses that call. For centuries, everywhere in the world there have been a multitude of demands competing for our attention resources, Bell said.

In defense of the Stanford study, however, Nass clarified that his research focused on media multitaskers—people who actively surf the web, use Facebook, listen to music, and engage in other activities—which he argues is different from physical multitasking. That's because with an all-in-one device like the iPhone, we are often jumping between tasks that are unrelated, like reading an e-mail during a meeting or checking Facebook while writing an essay. "When we're talking about smartphones we're talking about media multitasking, and that's where we're not built for that," Nass said. "By multitasking we mean doing unrelated media activities: If we're in the jungle we don't just look at a tree, we look around and look for tigers."[18]

In response, Bell argued that "physical multitasking" activities can be just as unrelated as media tasks, adding that scientific literature does not draw a fundamental distinction between physical multitasking and media multitasking. He also noted that a causality issue arises: whether heavy media usage causes these effects or whether people who are less able to exclusively focus and switch tasks prefer more media concurrently is unclear. Moreover, with the rectangle experiment, we could interpret the heavy multitaskers as being more distractable or simply that they have a wider net of attention and are more able to pick up peripheral information.

Despite his disagreements, however, Bell called Nass's study "excellent" and an important step into this area of research. "This is a valuable study because we need to start understanding how information technology affects us in our day-to-day life," he said. "We have precious few of these studies and we need more."[19]

Nonetheless, one of the most influential critiques of our 24/7 online culture remains Carr's book *The Shallows*, in which he argues that the Internet is damaging our brains by overloading us with data and ultimately rewiring people into shallow thinkers.

An Overloaded Brain?

In 2007 UCLA professor Gary Small gathered a group of six volunteers, three experienced web surfers and three novices, to study the Internet's effects on brain activity.[20] For the experiment, each volunteer wore a pair of goggles that projected web pages, and one by one each volunteer slid into the cylinder of a whole-brain magnetic resonance imager while Small instructed them to surf the Internet.

Small observed that brain activity for the experienced web junkies was considerably more extensive than that of the rookies, especially in the prefrontal cortex area. He then projected normal blocks of text on the subjects' goggles and saw no significant

difference between the brain activity of the two groups. However, just six days later Small repeated the tests and saw that the
beginner web surfers, who had agreed to spend an hour each
day searching the web, were already showing dramatic changes
in brain activity. The Internet, Small concluded, is "rapidly and
profoundly altering our brains."

Carr cites Smalls' study to illustrate the point that the Internet
is reshaping our brains in a bad way. In one part of his book Carr
focuses on the cognitive effects of hypertext—linking on the Internet—citing studies that found that Internet browsing causes
cognitive overload, which disrupts concentration. As a result,
the Internet as a whole, Carr argues, weakens our comprehension and transforms us into shallower thinkers:

> We know that the human brain is highly plastic; neurons and
> synapses change as circumstances change. When we adapt to a new
> cultural phenomenon, including the use of a new medium, we end
> up with a different brain, says Michael Merzenich, a pioneer of the
> field of neuroplasticity. That means our online habits continue to
> reverberate in the workings of our brain cells even when we're not
> at a computer. We're exercising the neural circuits devoted to skim-
> ming and multitasking while ignoring those used for reading and
> thinking deeply.[21]

Carr cites additional studies that suggest hyperlinking causes
cognitive overload. In a 2001 study two Canadian scholars asked
seventy people to read "The Demon Lover," a short story by
Elizabeth Bowen.[22] The subjects were split into two groups: one
group read the story in a traditional linear-text format, and at the
end of each passage they would click a "Next" button to move
ahead; the other group read a version that required them to click

on highlighted words to move ahead. The study found that the hypertext readers took longer to read the document, and they were seven times more likely to find it confusing. "Dazzled by the Net's treasures, we are blind to the damage we may be doing to our intellectual lives and even our culture," Carr writes.[23]

However, both Bell and Nass agree that claiming that the brain has been "damaged" as a result of technology usage is a dangerous, inaccurate statement. The idea of neuroplasticity— the brain's ability to change—has been a popularly abused term. Only gross changes seen in the organ, such as obvious tissue lesion or atrophy, can demonstrate a truly "damaging" effect on the brain, Bell said. Even if brain activity is in fact changing during Internet usage, this does not mean much: the brain is supposed to change in every given moment of every day because that's precisely what the brain does.

Furthermore, there is no definitive proof that cognitive overload is permanent, affecting us even when we step away from the computer. The aforementioned tests involved people who were being monitored while browsing the Internet, not while they were taking a stroll through the woods. "Neuroplasticity is just what the brain does," Bell said. "It's like a reporter from a crime scene saying there was 'movement' during the incident. We have learned nothing we didn't already know."[24]

Secondly, another consideration to keep in mind is that, increasingly, web browsing is an "older" technology. Hypertext is merely an interface element of a traditional web browser—an interface that's being supplanted. Newer always-connected mobile devices, such as iPhones, Android smartphones, and Apple's new iPad tablet, have introduced a smarter interface, thereby mitigating Internet distractions and making data more usable. These newer devices still include web browsers and, of course, web pages with hypertext, but they present content in a way completely different from a personal computer. Because these gadgets are so recent, we will have to wait for researchers to investigate their effects on the human brain; nonetheless, already we can draw some conclusions from previous studies.

In 2004 Muhammet Demirbilek, a post-doc researcher, conducted a study on 150 students at the University of Florida to examine the effects of different computer-window interfaces on learning.[25] He compared two interfaces—a tiled-windows interface, in which windows were displayed next to each other in their entirety, versus an overlapping-windows interface, in which windows were laid on top of each other like a stack of paper.

Inside a computer lab, the participants were randomly assigned to work in either the tiled-windows interface mode or the overlapping-windows mode. Each mode contained a multimedia learning environment that required the students to complete certain tasks. Demirbilek focused on measuring the students' disorientation and cognitive overload.

To measure disorientation, Demirbilek recorded each student's window use by keeping track of the number of informational "nodes" that were accessed to complete each task—in other words, the number of steps each user took before finishing an activity. For each task, based on the number of nodes a student accessed, Demirbilek labeled users either "oriented" or "completely lost" in the hypermedia system. To measure cognitive overload, he timed the students to see how long they took to react to different interactions. For instance, in one part of the study the participants were required to click a button as soon as the background color of a window changed.

After completing his study, Demirbilek found that subjects using the tiled-windows interface were significantly less disoriented than subjects using an overlapping-windows interface. He also found that participants in the tiled-windows environment suffered significantly less from cognitive overload than those working with overlapping windows.

In conclusion, students using the tiled-windows interface were able to find specific information more easily and engage with it more deeply, whereas students working with overlapping windows struggled to see how parts of a knowledge base were related and often omitted large pieces of information. In short, students using the tiled-windows interface were able to learn considerably better than those working with overlapping windows. "The tiled-windows interface treatment provided help to users, enabling them to efficiently communicate with the hypermedia learning environment," Demirbilek wrote.

Demirbilek's conclusions don't contradict Carr's assertions, but they do suggest that the gap where information is lost between short-and long-term memory is not due solely to hyperlinking but also to the disorienting nature of the interface used. According to Demirbilek's study, Carr is correct that the traditional PC-computing environment (such as Windows or Mac OS X), which uses an overlapping-windows interface, may be conducive to shallower learning.

But is that conclusion still valid in today's digital world? Apple's iOS operating system, powering both the iPhone and the iPad, introduced an interface that abandoned the traditional graphical user interface. Gone are the mouse pointer and cascading windows; now, the only pointer is the human finger that controls a multitouch screen. More importantly, every app launched completely fills the screen; we can view only one app at a time. Competing smartphones also assume the same type of full-screen interface, and going forward, we can expect future tablet computers competing with the iPad to replicate the single-screen interface.

As the number of tablet computer users continues to grow,* more web developers will feel pressured to scrap the busy website interfaces we are currently accustomed to.[26] Gone will be the distracting pop-up windows that Demirbilek found to be causing cognitive overload. Puny, squint-inducing boxes will be replaced with large, touchable icons. Perhaps the iPad is poised to improve user orientation and learning. "The interface of iPad could work well for us," Demirbilek told me. "We use our hands instead of a keyboard or mouse, and it fits exactly how we behave and think in real life. In addition, the iPad interface looks easier for us because it has larger-size text and bigger icons. It is less likely to cause cognitive overload to the user, based on my studies."[27]

By displaying only one app or one piece of content at a time, the iPad will help us focus and, in turn, absorb information more easily. As we can imagine, however, that solves one problem while creating another. In 1995 researchers at the University of Minnesota conducted a study on tiled windows during which they found that fourth-grade students benefited from tiled-windows interfaces compared to overlapping windows because multiple windows, displayed in their entirety, assisted in completing tasks when more than one source of information is needed to solve a problem.[28] With this in mind, although the iPad reduces elements of distraction and potentially enhances user orientation, its lack of windows also eliminates the ability to read information from multiple sources simultaneously on a single screen.

However, a single-screen interface doesn't altogether remove the ability to read information from multiple sources; one simply

---

*Apple sold almost fifteen million iPads in 2010.

has to read the sources one at a time. This shortcoming makes the iPad lack as a productivity device for doing work, but already Apple is addresses this limitation with a software feature called "Fast App Switching," which saves closed apps in a suspended hibernation state so a user can quickly resume where she left off when she reopens each app. Apple will address further shortcomings over time as the company continues to iterate the operating system.

This research suggests that the single-screen iOS interface is more conducive to slower but deeper learning than the overlapping-windowed PC interface preceding it. If so, the upcoming crop of always-on gadgets is poised to help us concentrate and learn.

We are unlikely to reach a solid consensus regarding the impact of an always-on lifestyle anytime soon, perhaps because the Internet and "media" are so broadly encompassing and there are so many different ways to participate. Some data show that gamers are better at absorbing and reacting to information than nongamers, but other evidence suggests that television viewing has a negative impact on teenagers' ability to concentrate. We can't simply bundle conclusions of various media studies.

The impact on brain activity seems to be an unresolved issue when evaluating what being always on means. Perhaps examining always-on's impact on social and behavioral activity would be more useful. Do we feel worse? Perform worse? Act worse as a result? Do Internet media have positive effects on our well-being?

Chapter 7

for
better
or worse

Kim Jae-beom and Yun-jeong were meant for each other. They met in 2008 in an online chat room, fell in love, and soon were married. They each had a daughter around the same age, and the four of them would venture through foreign cities, meeting new people and sharing magical experiences. About a year later they had a child together—a daughter they named Sa-rang, or "love" in Korean. She was born prematurely, a bit over six pounds. She weighed about a pound lighter at three months old, when the coroner declared her dead of malnutrition.

After half a year on the run, the Kims were arrested in March 2010 for negligent homicide, accused of starving Sa-rang to death. Prosecutors said they had fed the baby milk only two or three times a day, before and after ten-hour sessions of dwelling in an Internet cafe. Rather than caring attentively for their baby, the Kims were busy living second lives in Prius Online, a game in which they raised virtual daughters called "Anima" to assist them on their journeys through a fantasy world slaying monsters and completing missions. After a twelve-hour game session on the morning of September 24, 2009, the Kims came home to find Sa-rang dead, "eyes open and her ribs showing," according to their lawyer. "I am sorry for being such a bad mother to my baby," Mrs. Kim said, sobbing, during the couple's trial. The Kims were sentenced to two years in prison, though the sentencing was delayed for Yun-jeong because she was already seven months pregnant with another child.[1]

The death of Sa-rang struck an emotional chord throughout South Korea, arguably the most wired society in the world, where citizens take video gaming as seriously as we do basketball in the states. The Kims' incident, along with other similar episodes, helped legitimize widespread concern about Internet media's potential impact on mental and physical health, thus prompting the South Korean government to open rehabilitation centers to treat online addicts as if the Internet were a chemical drug.

Following the Kims' story, attention-grabbing headlines continued to fuel a negative stereotype surrounding the always-on culture. In February 2010 a twenty-two-year-old Korean man was arrested for stabbing his mother to death because she nagged at him about his video game addiction. Six months later a Hawaiian man sued a software company, alleging that their game Lineage II made him so addicted that he was "unable to function independently in usual daily activities."[2]

Pessimism toward technology seems to be even more fashionable today now that smartphones have become powerful enough to carry all sorts of potentially addictive experiences like games, live TV programming, and social networking services. It is easy to assume that smartphones are powerful enablers of unhealthy behavior for people with techno-addictive tendencies, now that we can carry an anything-anytime-anywhere experience in our pockets. This is like carrying around a digital crack pipe.

But as tempting as pointing fingers at a new technology for affecting the Kims and others may be, there are legitimate questions to ask. Did Internet-connected gadgets and media condition the Kims to lose control and ultimately starve their baby to death, or were they already depressed individuals in search of an escape, which they found in a virtual world where problems were settled with a few clicks of a mouse? Was Lineage II really so addictive that it deprived the Hawaii man of his ability to wake up on time, take a bath, and socialize with friends, or did he already have an addictive personality that preconditioned him to become so obsessed with the game?

All the various sides of the debate pushed aside, one thing is certain: we seem to have stumbled into a rabbit hole of self-reflection, with more and more of us pondering about how new technology is changing our social behavior. Just how are we changing, and are these changes for better or worse?

This is a complex question that can only be addressed by examining always-on's effects on the different aspects of everyday life, such as mental health, workplace productivity, and social behavior. Furthermore, examining the impact of each of the different types of content is important; the "Internet" and "technology" should not be lumped into a single homogenous experience. Attempting to generalize "the Internet" as good or bad is like saying "food" is good or bad; however, different types of food can be healthy or unhealthy depending on the amount one consumes. The answer to our question, then, as you might expect, isn't at all simple.

Game On

Video games were the focus of the aforementioned fear-mongering news stories—and games tend to be the biggest sellers in the software industry—so let's start there.

In media discussions video games take a tough beating as a "trash" medium, similar to the way comics are typically frowned on in the book industry. Influenced by news stories today, concerned parents and critics accuse video games of being evil brain mushers that condition players to assume the traits of characters they pretend to be: violent warriors, blood-shedding soldiers, and so on.

So are video games really hurting our brains and turning us into uncivilized creatures? Just how would we determine that? Robert Weis and Brittany Cerankosky of Denison University had a clever idea: hand a game console to boys who don't own video games and watch what happens to their academic performance after four months. The psychologists found that the boys who received a game system immediately spent more time gaming and less time in after-school academic and social activities than children who weren't given consoles.[3] As we might guess, the boys who spent more time gaming immediately started scoring lower on reading and writing tests, and teachers were more likely to report problems about how these newbie gamers were doing in class.

At first glance the results of the Denison University study look as if gaming is a proven culprit for diminished brain power and interest in school activities. However, the experiment doesn't demonstrate a correlation between video games and the harm to kids' brains and personalities. The key observation was that the gamers tended to engage less in after-school academic activities like study sessions and tutoring. So although video games were the distraction that pulled these kids away from studying, which in turn led to their decreased academic performance, they didn't cause any psychological harm.

As a matter of fact, video games have been shown to boost our brain power in some respects. A large number of studies have found that regular action-gamers are superior at absorbing contextual information than nongamers. A neuroimaging study on Tetris gamers found improved cortical efficiency (improved performance, reduced metabolic load) after a practice period.[4] Further, a separate study found that people who regularly play action video games have reduced reaction times in attention tasks without compromising accuracy or increasing impulsivity.[5]

Drilling deeper into psychological studies, changes to behavior are clearly more tied to the type of content rather than the type of media. To demonstrate, UC San Francisco psychologist Sonya Brady conducted a study on media violence, surveying participants' attitudes on interpersonal violence, criminal justice policies, and military activities. She found that the main factor driving aggressive attitude was sports. Avid watchers of contact sports (such as football) and people who regularly play sports-themed video games were significantly more likely to have favorable attitudes toward military-sanctioned violence and punitive criminal justice policies.[6] The question of games' psychological effects all boils down to content, not the medium.

Still, the video game medium, like most new media technologies, has a negative stigma attached to it, perhaps because of the unattractive image of people glued to screens all day while banging away at joystick buttons. And despite Brady's findings about athletic sports fostering more aggressive personalities, we don't see news stories about a bloody murderer who happened to love football leading to discussions about football being bad for the body and soul. In our eyes, an outdoor activity that involves lots of fresh air and cardiovascular exercise can't be "bad."

This is not to say there shouldn't be any concern about gaming. The common thread among stories about video game deaths is the element of addiction. And we must admit that video games can be very addictive. Everyone at every age seems to have grown obsessed with some sort of game at some point. Many people know some mothers hooked on Bejeweled, a father who plays Full Tilt Poker, and a friend who clicks around in Star-Craft.

Scientists have repeatedly found, based on a measure of hours they spent gaming and some psychosocial variables to determine pathological behavior (pathology is defined as an individual's inability to function in society), that a significant number of youth—about 9 percent, according to statistics—are "addicted."[7] Another 10 to 20 percent are engaging in "at-risk" behavior. Because of these findings, many psychologists take game addiction seriously.

Research has shown that gaming experiences activate the same part of the brain that has been strongly implicated in drug addiction and behavioral disorders—the frontostriatal pathway, where circuits help us control our voluntary behaviors. This is also the same part of the brain that's activated by plenty of things we find pleasurable, like drinking soda, listening to music or receiving a compliment. However, games are especially stimulating because the human body can respond to them as if they were a real, physical experience. High-motion images are more potent than words or sounds, so when we see an object, our nerves very quickly send images to our brains to compare what we see to basic shapes stored in our emotional memory. As a result, no mechanical part of the brain reminds the gamer that she's having a media experience versus a physical one.[8] Consequently, games can be very addictive because they present experiences that are so closely related to our everyday lives; after all, balancing our lives in physical reality in itself is difficult enough already. Furthermore, gaming addiction is complicated because nothing physically changes the brain immediately; the addiction is behavioral. "Almost nobody understands enough about games or themselves to keep their play balanced," say Neils Clark and P. Shavaun Scott in their book *Game Addiction*.[9]

Beyond games, a growing anxiety centers on the always-on culture of social networking. Are people who actively use social networking sites escaping or replacing their reality, or are they using these tools to deepen their relationship with the real world?

Mirror, Mirror on the Wall

We hear it time and time again: "We're living in the age of Generation Me," the age in which teens are more self-obsessed than ever before. Even if that's true, "self-obsession" is not necessarily a bad thing: a consensus of studies on social networking finds that digital mirror on the wall to be a huge benefit for someone's social well-being. Multiple studies show that Facebook increases one's self-esteem because it provides reassurance of where one is in life in relation to everyone else. Further, beyond simply making someone feel better, the benefits of Facebook get even more interesting.

In fact, those Facebook users who collect friends like stamps and post a status update every other minute are more likely to persevere through college than less active users. A study led by Abilene Christian University followed the Facebook profiles of 375 first-semester freshman students for nine months to examine how Facebook activity can be used as a predictor for a student's likelihood to stay in school. The research found that students who returned to school after freshman year had significantly more Facebook friends and wall posts than those who didn't return.[10] "The study was able to show that these students who are more active on Facebook are also out there getting involved, making new friends and taking part of activities that the university provides for them," said Jason Morris, an assistant professor of education and director of higher education at Abilene Christian, who authored the research article.

The study, only recently published by the *Journal of College Student Retention,* focused on students from fall of 2006 to summer of 2007. In it, students who opted to continue on to their sophomore year had on average twenty-seven more friends and fifty-nine more wall posts than those who dropped out, according to the study.

For other variables, Abilene Christian measured the number of Facebook groups joined and photo albums posted by the students, but in that regard, the statistical differences were negligible. Researchers determined that wall posts and the number of Facebook friends were the most significant predictors for determining a user's Facebook activeness, which in turn reflected their enthusiasm for the academic world surrounding them.

Abilene Christian's study emerges at an interesting time, when researchers and technologists are debating whether technologies such as social networking sites and smartphones are bringing people together or isolating them. In another recent study, based on surveys measuring empathy among almost fourteen thousand college students over the last thirty years, University of Michigan researchers found that today's college students are significantly less empathetic than college students of the 1980s and 1990s.[11] The researchers theorized that the drop in empathy might be due to students' excessive exposure to media, such as violent video games, which "numbs people to the pain of others." They also suggested that perhaps connecting with friends online makes shutting out real-world issues easier.

"The ease of having 'friends' online might make people more likely to just tune out when they don't feel like responding to others' problems, a behavior that could carry over offline," said Edward O'Brien, a University of Michigan graduate student who helped with the study.

However, Abilene Christian's Facebook study led researchers to different interpretations. They believe that rather than sites like Facebook helping people escape reality, they are actually a mirror for real-life interactions among the participants. The students who were more actively connecting with people on Facebook were most likely already connectors in the real world. "At the time we did this study, the big debate was whether or not the Facebook world was a virtual pseudo social world or whether or not it actually reflected real world relationships," said Richard Beck, associate professor and chair of psychology at Abilene Christian, who came up with the idea of the Facebook study. "[The study] seemed to indicate that what was going on Facebook was paralleling their social experience on campus. Rather than replacing it, it was mirroring it."

Abilene Christian's study is also an example of how social networking sites—as invasive to our privacy as they may be—can potentially serve as a more objective window into human behavior than surveys, Morris said. It is worth noting that some studies have shown that self-reported results, like the empathy survey that the University of Michigan conducted, have resulted in heavy bias and, at times, inaccurate results. "Instead of using [students'] perceptions, we remeasured . . . with actual behaviors, which makes that a little more powerful as a study," Morris said.

Beyond gawking at people's Facebook profiles, some psychologists are studying the impacts of social networks and cell phones on writing—and what this can tell us about our brains.

A Word's Worth

When we were kids, teachers occasionally refused to let us use calculators during math tests for fear that we would never learn to compute numbers with our own brains if we always relied on computers. Naturally, there is a similar fear surrounding the use of all-in-one smartphones: we will forget to do all sorts of things that software applications can handle. In China and Japan especially there is a growing concern about people's diminishing ability to write. The *China Youth Daily* commissioned a poll that found that 83 percent of the 2,072 respondents admitted to having problems writing Chinese characters, and this is presumably because the constant use of keyboard-equipped computers and smartphones was making them forget the strokes learned through handwriting. The same survey found that young Japanese people were having the same problem memorizing words, a phenomenon it called "character amnesia."[12]

Japanese and Chinese smartphone users type words in Pinyin, a phonetic system that uses the English alphabet and accent marks to represent the pronunciation of Chinese words, and software then automatically translates Pinyin into Chinese or Japanese characters. People presume that this automatic translator is the cause of this so-called character amnesia.

But without further data and experimentation, we simply can't draw this conclusion. Instead, we should ask ourselves, Is technology really making the Chinese and Japanese forget how to write characters, or are they learning more words now than they ever were before and are therefore struggling to remember how to write everything? Perhaps the hint of an answer can be found in a late 2010 study on Pinyin, which tested Chinese kindergarteners on their ability to read and pronounce Chinese, both in its traditional character form as well as Pinyin.[13] The test found that when writing in Pinyin, the children could write more words than they could write traditional characters. Most importantly, the research found that the kids demonstrated an impressively accurate understanding of the pronunciation and accents of characters when they transcribed them into Pinyin. Because phonetics are an extremely important part of speaking Chinese, the study's authors believe that over time Pinyin could be a strong aid to complement reading traditional Chinese characters. Although this study doesn't ultimately disprove the thesis that smartphones are making the Japanese and Chinese forget how to read, it does demonstrate the possibility of a vastly different conclusion.

Here in the United States, researchers are also eyeing always-on's impacts on the English language. After all, teenagers armed with keyboard-equipped cell phones and laptops are writing more than ever before. Working with this premise, the Pew Research Center in 2008 conducted a study focusing on academic writing, and it received mixed results.[14] Pew saw some stylistic degradation in teenagers' academic writing: based on phone interviews conducted with seven hundred twelve- to seventeen-year-olds and their parents all around the nation, 50 percent of teens confessed that they sometimes use informal writing styles, such as improper capitalization and punctuation, in their school assignments; 38 percent admitted they use text shortcuts such as "LOL" (laugh out loud); and 25 percent said they use emoticons, which are symbols such as smiley faces :-).

With all that said, however, most interestingly, 60 percent of the students responding to the Pew study believed that the written communication taking place on the Internet is not actual writing. "The main reason teens use the Internet and cell phones is to exploit their communication features. Yet despite the nearly ubiquitous use of these tools by teens, they see an important distinction between the 'writing' they do for school and outside of school for personal reasons, and the 'communication' they enjoy via instant messaging, phone text messaging, email and social networking sites," Pew wrote. These statistics suggest that the Internet colloquialisms and mannerisms slipping into school assignments are a voluntary act, which means that online chatting isn't subconsciously morphing teenagers to write worse. Therefore, regular online communication may not negatively affect writing style after all.

In fact, writing underperformance may be diminishing among always-on teens. The Nation's Report Card, an institute that aggregates data on grades among elementary and secondary students, did not find significant improvements in writing proficiency between the year 2002 and the year 2007.[15] However, during that time period the firm did see a substantial decrease in the number of students writing below basic levels of competence: in 2007 13 percent of eighth graders and 18 percent of twelfth graders scored below a basic level of writing proficiency, down from the 15 percent of eighth graders and 26 percent of twelfth graders in 2002 writing below the basic level. These numbers suggest that even if the Internet isn't making more students write at higher skill levels, it might at least be making more kids average.

To go beyond writing, however, the question of whether or not a wired lifestyle is actually making students perform better in school remains open. By 2005 almost all public schools had access to the Internet, and the *Digest of Education Statistics* did not find a sanctifying trend between education and technology.[16] Students' nationwide scores in 2009 on tests of reading, writing, science, and math were collectively both up and down, depending on the subject and the grade being tested. Moreover, the Pew study also found that teens who own more technology tools do not write more in their free time than do less gadget-rich teens. In other words, kids who are going to write are going to write with or without tech, and those who aren't going to write, aren't.

Ultimately, performance data for teenagers' writing did not lead to solid conclusions about their overall academic performance in the age of always on. However, one part of Pew's study on writing is rather intriguing: kids don't consider electronic communication to be writing, even though they actually are writing words. Perhaps they are on to something.

A word today is worth much more than it was just ten years ago: a single micro message published on Twitter, for example, can throw an entire nation in upheaval (e.g., the Twitter-driven protests in Iran). As such, the definition and role of writing is quickly evolving. Before the printing press, access to the written word was considered a privilege because reproductions of writings were limited; academic learning was significantly more relational, as mentors guided small groups of students. Then, the printing press expedited the mass-production of knowledge and drove education to the textbook model. Today, students are seeing the written word come full circle—not simply a means of information containment and dissemination but also transaction. Our process of learning is becoming interactive and relational again as a result of the Internet, thereby renewing the old mentor-and-student relationship that eroded as we came to rely on textbooks.

That we are not seeing any significant improvement or deterioration in academic performance is hardly surprising; after all, the educational system as a whole remains fundamentally unchanged despite the rampant progress of technology. Most teachers still live religiously by the textbook model, and shrinking state budgets stifle efforts to innovate education.

Imagine if, rather than going home, reading textbooks, and filling out repetitive assignments, teachers encouraged students to engage in online exercises to learn to speak new languages or play different musical instruments or to write a historical research paper collaboratively with another teen from the other side of the world—that is, imagine if educators treated learning as more of an online, interactive experience, similar to how Magnus Carlsen viewed his own virtual training to become the world's best chess player.

If teachers and administrators substantially revised general classroom curricula so as to integrate the web to complement the textbook we would likely see some interesting results. Judging from the fact that fewer students are writing below basic proficiency levels today compared to only a few years ago, we would likely see a more significant improvement in general academic performance if educators were more plugged in. Learning could become more like a video game. In fact, author James Paul Gee challenges educators to think beyond traditional literacy to consider that the interactive environments inside video games could be well suited for learning and literacy. "The content of video games, when they are played actively and critically is something like this: They situate meaning in a multimodal space through embodied experiences to solve problems and reflect on the intricacies of the design of imagined worlds and the design of both real and imagined social relationships and identities in the modern world."[17]

In his book *What Video Games Have to Teach Us about Learning and Literacy,* Gee points out that language is much more than simply words: graphics, symbols, body language, and color all take on importance as well, and we call this "semiotic grammar."[18] Images incorporated into textbooks are semiotic elements that can potentially enhance a person's comprehension of a word's meaning. An interactive video game-like environment could take semiotic learning several steps further. In short, Gee argues that a video game environment can train children to engage with the real world by injecting them into a virtual one in which they can learn by doing.

Again, however, because academic structure remains stuck with a textbook model, we are not likely to see educators embrace video games as a learning tool anytime soon. But we can understand more about always-on's impacts on "learning by doing" by exploring its effects in the workplace.

Wired Workers

When I met Tim Ferriss at a tea lounge in San Francisco, he was the only man wearing shorts and flip flops. We were at a networking event for local entrepreneurs, and among the men and women flaunting power suits, Ferriss's casual attire struck me as odd. I understood his getup once he introduced himself to me as the author of *The 4-Hour Work Week,* a motivational book that instructs its readers on how to become rich by working only four hours each week.[19]

Hearing about his book, I thought it was ludicrous. But reading it later, I found some of the ideas intriguing. Although I doubt anyone can easily achieve the goal of becoming wealthy by working only a few hours each week, Ferriss makes some radical suggestions about how we do work in general. He preaches drastically cutting down on time spent e-mailing and completely removing any form of digital time wasting such as Twitter or Facebook. His idea is to snip out any unnecessary distractions so that people and their colleagues can get their work done in fewer hours, go home, and enjoy life a little longer. "Never check e-mail first thing in the morning," Ferriss writes. "Instead, complete your most important task before 11:00 a.m. to avoid using lunch or reading e-mail as a postponement excuse."[20]

In other words, in an era when many of us are always connected, Ferriss suggests being selectively connected in an extreme way. Though his book is purely unscientific and bereft of actual research, he raises some interesting questions, most notably, Is technology empowering us as workers, or is it crippling us with endless distractions?

Ferriss contends that distractions are paralyzing us, and according to recent research, he may be right. Interruption scientist Gloria Mark led a study that shadowed twenty-four tech-rich knowledge workers over a thirteen-month span.[21] The researchers observed the workers and jotted down the time (to the second) whenever the subjects switched tasks—everything from opening a document to composing an e-mail, and from making a phone call to talking to surrounding people. At the end of every day a researcher interviewed each subject for clarification on their activities.

The numbers in Mark's study were striking—on average workers spent only eleven minutes on a project before switching to another, and while focusing on a project, they typically changed tasks every three minutes. Once distracted, they took about twenty-five minutes before returning to the original task.

Making similar observations, the business research firm Basex estimates that office interruptions and the requisite recovery time now consume 28 percent of a worker's day.[22] The firm calculated that altogether, distractions take up to 2.1 hours of an average knowledge worker's day, thus costing the US economy $588 billion a year, according to Basex.

Later Mark conducted a separate similar study and found that workers who were routinely interrupted were more apt to feel frustrated, pressured, and stressed. Furthermore, Pew Research conducted a survey that found that although 80 percent of "wired and ready" workers believe technology has improved their ability to do their job, nearly 50 percent of them say technology increases the number of hours they work as well as the stress level of their jobs.[23] Thus, technology may be empowering people to do more than ever before, but clearly it is also taxing workers' energy as well and costing businesses a lot of time.

However, prohibiting employees from freely browsing the web isn't necessarily the solution. Another study conducted by the University of Melbourne found that workers who engage in "leisure browsing" are more productive than those who do not.[24] "People who do surf the Internet for fun at work—within a reasonable limit of less than 20% of their total time in the office—are more productive by about 9% than those who don't," says Brent Coker, a University of Melbourne professor in the department of management and marketing. He explained that leisure browsing allows for workers to restore their concentration levels and, ultimately, do more work in a given day. However, he says that web browsing needs to be done in moderation, as 14 percent of Internet users in Australia show signs of Internet addiction, spending more than a normal amount of time online.

Additionally, the Pew Research Center surveyed workers and found that the vast majority of "wired and ready" workers note big improvements in their work lives and their ability to share ideas with coworkers due to the influence of technologies such as the Internet, e-mail, cell phones, and instant messaging.

In summary, hooking into the Internet doesn't necessarily make workers more productive, even though they do gain additional tools to be more capable. For the most part, wired workers are prone to distractions; only light Internet-usage workers were found to be more productive. Therefore, Ferriss's proposal for career success is feasible: the most successful workers will be those who can focus on relevant tasks and selectively ignore unnecessary distractions. This idea sounds familiar. Carlsen said he refuses to learn about his own intelligence out of fear that he will desire to learn too much and, consequently, lose his focus on chess. Apparently, both Carlsen and Ferriss agree that with the wealth of information available on the web, the "smartest" people are those who can stay focused on the most important information needed to achieve a goal.

Old Fears of New Media

You who are the father of letters, from a paternal love of your own children have been led to attribute to them a quality which they cannot have; for this discovery of yours will create forgetfulness in the learners' souls, because they will not use their memories; they will trust to the external written characters and not remember of themselves.

Such were the words of Socrates on, well, words. Centuries ago the philosopher warned that the act of writing would destroy our ability to memorize. Perhaps if he were alive today he would wish the above quote weren't written down—because he was wrong. Modern studies have shown that copying something down on paper actually is an act of repetition that helps us embed a thought into our memory.

Socrates, however, is not alone in his new media–phobic sentiments. Whenever a new technology dawns on us, at least a handful of cynics sound the alarms. Back in 1565 Swiss scientist Conrad Gessner authored a book criticizing the printed book, stating that information overload would overrun modern society. Then hundreds of years later, when books became standardized and schools were widely introduced, naysayers blasted education for being a risk to mental health. An 1883 article in the weekly medical journal the *Sanitarian* claimed that schools "exhaust the children's brains and nervous systems with complex and multiple studies, and ruin their bodies by protracted imprisonment." Neuropsychologist Vaughan Bell, who devoted an entire *Slate* column to retracing the history of media scares over then-new technologies, states, "The idea that new technology is 'over-loading us' in some way is as old as technology."[25]

Now the latest subject of fear is the smartphone nestled in our pockets, threatening to distract us with virtually any type of digital content we want—anytime and anywhere. We are repeating history, and just as we survived the wrath of the pen, the book, the radio, and the television, our minds and bodies will probably go on undamaged in an age that's always on. But that doesn't mean that being always on doesn't hurt.

Chapter 8

disconnected

When I was twelve, I opened a door to an alternate universe where I could be anyone I wanted to be. A professional skater, a nineteen-year-old named David from Wisconsin, even a cartoon character—I could change my identity by simply creating a screen name. I thought of the America Online chat rooms as grounds for a virtual Halloween that I could celebrate every day. I immersed myself in this mania. The sounds of a dial tone followed by the buzzes and hisses of a 56K modem became the soothing part of each day after school.

I was a tiny bit ahead of my time. None of my classmates then were familiar with online chatting. While my mother forced me to come home immediately after school each day, my classmates would walk to McDonald's for a 3 p.m. lunch, go over to each other's houses to play PlayStation, or drink someone's dad's beer. My family's computer became a friend who opened to me an entire social dimension.

Two years later my peers gradually began plugging in. They were signing on to America Online Instant Messenger (AIM) left

and right: PandaPoo54, BooGeR625, phlywitegy, TAN6ENT—
we each assumed our identities. We started with meaningless
small talk and sent each other our favorite hip-hop MP3s. But
then our interactions evolved. We began organizing our social
gatherings over AIM. Soon, we were having extensive conversations over AIM, ranging from the casual to the profound. My
phone barely rang anymore; the sporadic chime of an instant
message replaced it.

Chat didn't end there for me. Over the next ten years, in college and at each of my jobs, I would meet people physically first
and follow up with them through online chat to expand our relationships. Our online friendships took the front seat: we could
talk to each other anytime from anywhere with a web connection, so naturally we had more conversations on the Internet
than we did in real life.

I thrived in this connected lifestyle. In workplaces I politicked
through online chat, connecting with every important person I
needed to win over in order to climb the ladder rapidly. In my
romantic life I would meet women in person first and then flirt
with them online later; in fact, most of my relationships were
rooted in online conversations. I felt profoundly connected to so
many people. I thought this was the way to go, and anybody who
chose to be unplugged was living in an inferior reality.

But when I met Kristen, everything changed.

I was twenty-two at the time, fresh out of college, new to San
Francisco, and a little lost, tender, and lonely. That year I somehow ended up volunteering at the International Film Festival
during Cinco de Mayo as a ticket taker at the will-call table. A
small Asian woman with glasses and long black hair showed up
fifteen minutes late for our shift and sat down next to me.

"Hey, sorry I'm late," she said, nearly out of breath.

"No worries," I replied, without completely turning to look her in the eye. (The truth is that I was annoyed. I'm never late, and people who aren't on time annoy me.)

We exchanged a few words while taking people's tickets, and when the line disappeared during a screening, we introduced ourselves more formally. Kristen had recently quit her job as a dietician to go to grad school to study social psychology; I was a magazine editor at *Macworld*. This intrigued her: her ex-boyfriend had written an iPod app that our magazine had mentioned. I was intrigued with her: she was studying a topic she found truly fascinating, a sociological analysis of the human condition. We connected immediately—and in a way I have never connected with anyone before: without the aid of an online chat window. For the rest of our five-hour shift we rambled about our ex-lovers, our families, technology, and movies.

And then she popped the inevitable question: "How old are you?"

"Twenty-two. You?"

"Twenty-nine."

I froze.

"Well, I'm happy for you," I said. I could hear my voice deepen to sound more mature. "You recognized you were stuck in a job you didn't like, and you got out of it. Most people never even do that, not even at your age."

After our shift ended, we waved good-bye to each other outside the theater and headed our separate ways. I didn't get her number. I was too intimidated—a twenty-nine- and a twenty-two-year-old? It would never work. Of course, I regretted this immediately.

However, with the power of Google, Kristen found my Live-Journal weeks later and posted a comment saying hello. We

wrote a few e-mails to each other and met again in July. She
didn't use IMing or computers much at all; she only had an old,
shoddy iBook for occasional web browsing and e-mails. As
friends, we went hiking, watched a ton of movies, and tried new
restaurants. We decided to go on a vacation together to Florida,
and that's when we shifted gears.

We kissed shamelessly on the beach of Miami, buried each
other in the sand, and splashed around in lukewarm water. In
bed at the Sheraton, Kristen laid her entire body on mine and
pressed her head against my chest while rubbing her hand over
my heart.

"It feels like we're on cloud nine," she said. "I've never felt
this way before with anyone."

"I feel the same way," I replied. "Like we're floating on a separate plane of existence."

"Do you really mean that?"

"Yeah."

The honeymoon phase ended a few months later when Kristen
started school again. She never signed on IM, and she stopped
sending the occasional e-mail while I was at work. She only pas-
sively used her cell phone: she picked it up only when she felt
like it, and to her, text messaging was a foreign idea. For the entire summer we had spent every day together—she had a suitcase
with her belongings in my bedroom—but when school started
again, she packed it up and moved back home; I only saw her
during her pockets of free time.

We were disconnected from each other, both physically and
virtually. This shift, to me—a person who is always on, always
reachable—was terrifying.

A few miscommunications and fights later, Kristen left me. The love was short-lived but intense, a gnarly love that threw my entire life into upheaval. As easily as I once created a new screen name on America Online, I had created a new identity to fit into a twenty-nine-year-old's life in order to make her believe in me—a fresh-out-of-college youngster who barely had anything figured out. I had fooled her, and I fooled myself into thinking I was a mature and autonomous grown-up.

I didn't know who I was anymore.

So I did something melodramatic. I subjected myself to a social experiment that I knew would be extremely uncomfortable: I vowed to cease online communication for three weeks. I wouldn't unplug from the Internet entirely—I still had to do work, after all—but all my conversations would have to be face-to-face or at least over the phone, including no more electronic-text communication. In a column published in *Macworld*, I would later document my experience of disconnecting from the world of online chat.

A few days into my experiment, I walked over to my coworker Greg's cubicle with my morning cup of tea to say hello.

"I was waiting for your screen name to pop up on my buddy list!" he said. "Do you have any idea how this affects our relationship? Or how this affects me? Why is it always about you, Brian? What about me?"

Greg was clearly joking, but he was right. I knew he was tired of dialing my extension just to ask if I could go on a smoke break. Incidentally, my cubicle neighbor Heather pointed out that I had started "thinking out loud" regularly, which must have been distracting for her. My friends Jenn and Matt, who hadn't spoken to each other in over a year, began IMing each other to fill in the "social void" my absence had created for them. My boss Jim

Galbraith couldn't tell when I left for my lunch break because, before, my screen name would carry such information with an Away message.

"Why are you putting yourself through this?" my roommate Peter asked.

"I'm tired of words," I said. "I want to hear people's voices, see them in the flesh. I want to analyze their body language and look them in the eyes—like old times. I want to reconnect."

"You're a damn fool."

Perhaps I was foolish for thinking I could revert. Cutting myself off from electronic communication made life unnecessarily difficult; the inability to chat online was extremely inconvenient. For example, my friend Amy and I took a vacation to Las Vegas for Christmas, and planning the trip over the phone was far more difficult than it would have been had we used e-mail instead. Further, even though prior to beginning my experiment I sent an e-mail to all my friends and coworkers forewarning them about my three-week hiatus, they continued to e-mail or text message me, so I would have to call them just to respond to simple questions or comments.

By refusing to participate in online chatting, I was inconveniencing others for the sake of a silly experiment. Even calling friends out of the blue seemed intrusive; I caught them off guard, so they had trouble thinking of what to say. I ceased electronic-text communication to conduct an experiment on myself, but ultimately, signing off affected others in a negative way that seemed selfish.

On top of that, I felt restless, constantly scrambling to find errands and activities to fill the silence. I felt alone and out of the loop. After those three weeks I concluded that there was no going back: I was destined to an always-on lifestyle because almost everybody was already so plugged in. Disconnecting was like going off the grid.

The experiment seemed stupid, undertaken during what many would call a "quarter-life crisis," but I'm not alone in my curiosity about the effects of unplugging. Beyond my miniature social experiment, a small number of studies are analyzing the effects of unplugging from data. Research subjects have reported similar experiences to mine: feelings of withdrawal, being ostracized, difficulty connecting with peers, and more.

Hooked on Data

University of Maryland led the most high-profile study to date, titled "Unplugged."[1] The school challenged two hundred students to completely give up media—no Twitter, Facebook, IM, web browsing, television, or anything—for twenty-four hours and then asked them to report their experiences. After their day offline, the students documented their thoughts in a collective blog, which in the end added up to a total of 111,000 words— about two novels worth of reflections. The majority of them compared disconnecting from media to withdrawal symptoms from an addictive chemical drug. "I clearly am addicted and the dependency is sickening," said one person in the study. "I feel like most people these days are in a similar situation, for between having a BlackBerry, a laptop, a television and an iPod, people have become unable to shed their media skin."

"Although I started the day feeling good, I noticed my mood started to change around noon. I started to feel isolated and lonely. I received several phone calls that I could not answer," wrote another student. "By 2 p.m. I began to feel the urgent need to check my e-mail, and even thought of a million ideas of why I had to. I felt like a person on a deserted island."

So maybe we really do have a problem with disconnecting; after all, people seem to enjoy virtual interaction with each other even in dangerous situations like driving a car. University of Kansas researchers recently polled a group of 348 students and found that the majority of them (83 percent) believed that texting while driving was unsafe—even more unsafe than talking on the phone while driving—but 98 percent of them admitted to doing it anyway.[2]

The study's eighty-nine-item questionnaire asked students to rate perceived risks of different reasons to text (initiating a text or replying to one) as well as texting during various driving conditions. Most interestingly, the study found that students rated driving on the highway to be intensely risky, but most drivers reported they were just as likely to initiate a text while driving on the highway as they would while driving in normal road conditions. In other words, they were able to convince themselves that road conditions were safer, which made texting justifiable.

"People know it's harmful and yet they keep doing it and they tell themselves they have to do it," said Paul Atchley, an associate professor of psychology who led the study. Atchley chalks this behavior up to the theory of cognitive dissonance: persuading yourself that a behavior is less risky by engaging in it—for example, smokers who convince themselves that the risk of smoking declines when they smoke or the perceived risk of drunk driving decreases when people drive drunk. We grow attached to the lifestyle of being always on, and we want to stay plugged in—even when doing so is probably a bad idea.

But why do people feel they have to text at all times? Social networking sites, coupled with a constant Internet connection everywhere we go, support the need for social beings to belong, and some research has shown that exclusion from social networks and text messaging can reduce that feeling of belonging, and this sets off pain signals in the brain.

Psychologists Anita Smith and Kipling Williams led a 2004 research study that focused on ostracism in the world of text messaging.[3] To conduct the study, two actors would get in a room with one participant. The two actors and the one participant sat next to each other in chairs arranged sixty centimeters apart in a triangular formation. During the meet and greet an instructor told the group that they would be text messaging each other. Then the actors left the participant alone in the room. On a table was a Nokia cell phone for the participant and instructions on how to use it. An instructor then told the group to text each other the question, "Do you smoke?" If a participant said he or she smoked, the two actors would say they didn't smoke; if the student said he or she didn't smoke, the actors would say they smoked.

Then the actors were either instructed to begin text messaging the participant casually or to stop messaging the student altogether. The students who were ignored were called the "Ostracism" group; the students who the actors communicated with casually were called the "Inclusion" group. This part of the test lasted eight minutes.

The end result? The students answered a questionnaire, and as might be expected, the Ostracism group reported significantly lower levels of self-esteem and feelings of belonging as well as higher levels of negativity and anger than those in the Inclusion group reported. Many of those ostracized even explained their theory for why they were ostracized: because they did or did not smoke (keep in mind the actors were told to say they were the opposite of however the student answered the smoking question). "It is remarkable that text message ostracism was found to be so aversive, considering . . . the participants did not know the [actors] and did not anticipate any further interaction with them," the psychologists wrote. They argue that the immediate reaction to ostracism is to feel pain without even thinking about it because "humans have evolved to detect even the slightest hint of ostracism and to experience it negatively, to warn that something must be done in order to be reincluded and prevent a threat to survival."

So many of us like to text a lot because once we stop getting messages, we start to hurt, and then we become scared and unsure of ourselves. "We're social organisms; there's so many mechanisms built into the brain that are designed for socialization," Atchley said. "The telecommunications industry has hit on something we're built to do."

Research suggests that in the future, young adults and children will be permanently plugged in to an always-on lifestyle. But does this make us dependent, socially crippled idiots? Recall the introduction of this book, when my cab driver argued that the Amish were the smartest people on the planet because they live wholesome, technology-free lives and retain intimate social connections. However, a closer look at the Amish reveals that the community isn't all that tech-free.

Life Hackers

In 2009 Kevin Kelly, the founder of *Wired*, stepped out of his comfort zone to pay an extensive visit to an Amish community in Pennsylvania.[4] The techno-visionary spent several days with the Amish in order to understand their off-the-grid lifestyle. His findings about technology were rather striking and contradicted the stereotype that the Amish are Luddites. In fact, Kelly described them as meticulous hackers. "In many ways the view of the Amish as old-fashioned Luddites is an urban myth," he said.[5]

The Amish, Kelly goes on to explain, are not a monolithic group. "Their practices vary from parish to parish," he says. Although the Amish are slow to adopt new technologies, they don't absolutely shun gadgets. Rather, the Amish hack their way around technology so as to integrate it into their lives ever so gradually and cautiously, thereby ensuring it doesn't undermine their religious beliefs and their communal tenacity.

Accepting a technology starts with an early group of "early adopters" trying out a new gizmo and assessing whether the pros outweigh the cons. Some Amish, for example, own cell phones. These "Alpha geeks" don't see much of a difference between a wireless phone in your pocket and a phone you use to place calls in a phone booth. The justification for a phone connection is to dial the local fire department, for example, or for people in the community to communicate if they don't live near each other. Still, many remain concerned about cell phones because they are so small that they can be easily hidden, which could pose issues for a group dedicated to discouraging individualism. Nonetheless, the Amish have been experimenting with and analyzing cell phone adoption for at least ten years, Kelly says.

The Amish also segregate technology they have in their homes and technology they have at work. Many still don't have electricity running into their own homes, and this means they don't have TV, Internet, or phones. However, a man who runs a furniture-milling factory, for example, will have power running into his shop. The power comes from a giant diesel generator attached to a large tank that stores compressed air; the engine burns petroleum fuel to power a compressor that fills the reservoir with power—so in a nutshell, energy created out of compressed air. The Amish call this pneumatic system "Amish electricity," which can also power appliances such as blenders, sewing machines, and washers or dryers.

Furthermore, the Amish seem to draw a clear distinction between ownership and using. They won't get a driver's license or purchase a car or buy insurance, but they will call a taxi or hire drivers to take them to and from work; some parishes forbid car ownership because drivers might leave the community to visit other towns. This distinction between ownership and using is what fascinates me most. By not owning cars or connecting to electricity and other utilities, the Amish distance themselves from the everyday problems we experience such as expensive bills and bureaucratic paperwork, and yet they still reap the benefits of these technologies. Furthermore, they are still able to integrate select technologies into their lives in a nonintrusive way that is strictly for business or utility. "Amish lives are anything but anti-technological. In fact, on my several visits with them, I have found them to be ingenious hackers and tinkerers, the ultimate makers and do-it-yourselfers and surprisingly pro-technology," Kelly says. Overall, the Amish have effectively hacked their lifestyles so that technology presents solutions rather than problems.

The Amish don't exclude the outside world. In fact, in their tradition rumspringa ("running around"), beginning at age sixteen, adolescents gain the privilege to live in our always-on world for a few months or years, released from the rules of the church. The tradition holds that regardless of what a person does during rumspringa, they still have the choice to stick with the "out-sider" society or return to the Amish church and be baptized. If they return, however, they must stick to the rules. An estimated 85 to 90 percent of Amish teenagers return to the church.

That statistic is ambiguous. It could be attributed to the fact that some Amish can't deal with the cultural shock during rum-sringa, or perhaps they prefer the relationships they have fostered in their own community. Whatever the case may be, however, the Amish, from an outsider's perspective, seem to have more intimate, closely knit relationships with each other; the human spirit and a concerted effort to stick together is mostly what powers their technology. Though Kelly does point out that they are not completely self-reliant: "They do not mine the metal they build their mowers from. They do not drill or process the kerosene they use. They don't manufacture the solar panels on their roofs," he says, and so on. In other words, if everyone chose to be Amish, our society wouldn't be sustainable because no one would manufacture all the stuff we need—not even the bare minimum.

Kelly believes the ultimate trade-off for Amish contentment is revelation. He quotes a 1950-dated writing from sociologist David Riesman: "The more advanced the technology, on the whole, the more possible it is for a considerable number of human beings to imagine being somebody else." In regard to the Amish, this implies that by minimizing technology integration in their personal lives, they stifle themselves and, indirectly, the rest of the world around them. "They have not discovered, and cannot discover, who they can become," Kelly contends. But isn't that precisely the point of living off the grid—to contribute to the community you're close to rather than society as a whole, to cultivate your own garden rather than everybody else's?

Long ago I stopped creating new screen names and pretending to be other people in chat rooms like I did when I was a teen. As I have grown older, I have communicated with thousands of people, subtly tailoring my personality to connect with each intimately. I wear different masks in various contexts: on Twitter, I have accrued 12,000 followers who see my slightly professional, public persona. On Facebook with only my friends, I present myself as the sardonic thrill seeker they often know me to be.

But perhaps with Kristen I took altering my identity too far. Through our e-mails and real-life conversations, I altered my lifestyle and appearance to be someone I wasn't in order to be with her. So when she left me, I was in utter disarray about my identity—perhaps because I'm not simply one person but rather an expanding catalog of a human psyche, torn asunder by the chaos of the web. Perhaps there are many people younger than me, growing up in our vastly connected culture, who are very much the same way: complex and socially empowered but chaotic and unstable. Always-on media may deliver many psychological benefits, as I have outlined in earlier chapters, but it is also a powerful drug that should make us wary of how we behave when we have access to so much data everywhere we go.

Completely unplugging doesn't seem to be an option. If we were to experiment with an Amish lifestyle—a reverse rumspringa, if you will—it would probably result in 90 percent of us returning to our plugged-in homes. But I think what we can learn from the Amish and their minimal lifestyle is worth meditating on. The Amish carefully assess whether a technology presents benefits or harm; as a result, they ban technologies more often than they adopt them. Perhaps in a connected society, we can evaluate how to use our gadgets in a healthy and con-

structive way as well as how to combat behaviors that are negatively affecting our health and relationships.

A more pragmatic solution than unplugging could be setting
our own rules to gain some control. An architect I chatted with
at a party proposed the clever idea that people should set apart a
room in their homes where technology is completely forbidden
so they can rest or interact with each other without digital distraction. "No cell phones in this room" would be the new equivalent of "Please take off your shoes before entering." Provided
one has the space, this seems like an idea worth trying out—and
it is a construction concept that architects should consider in-
corporating when designing new houses.

There are plenty of us with compulsive tendencies that smartphones are enabling, as depicted by the text messaging study. A
lot of my friends in Silicon Valley have a habit of whipping out
their smartphone every other minute, even while I am with them
physically, and this habit makes me feel ostracized and irritated,
although I know they don't intend to make me feel that way.
Technology manufacturers should be sensitive to our compulsive tendencies and start building features to help us regulate usage: Apple, Google, and Microsoft, for example, should begin
shipping phones with a tool that allows us to ban checking text
messages or receiving calls or e-mails at dinner time. In the
meantime we should be cognizant of how our technology habits
can displease people we care about: when you are with a group,
think about turning off your phone in order to stay mentally with
each other in the room.

And then there are our children. Forbidding them from using
computers or cell phones, even at a really young age, would
probably be devastatingly crippling at this point in our society's
technological progression. Nonetheless, parents should do their

part by encouraging their kids to go outside and play—but again, I think also manufacturers should be more accommodating to our children's needs. They should begin shipping affordable devices for children with lightweight features and perhaps some educational software and games.* Some sophisticated parental controls could easily be made available in these for-teen phones to allow guardians to regulate usage: no porn, of course, and no surpassing a certain number of text messages or browsing the web after bedtime, to name a few examples. Current mainstream smartphones feature poor parental controls that can't closely monitor or regulate children's activities. (The iPhone, especially, has a poor parental-control feature that offers barely any flexibility, only allowing parents to flip features on or off instead of being able to limit usage to certain durations or types of content.) Going forward, children will need always-on data in order to excel, but we also should demand from manufacturers control over how they access and use data.

If there is one thing we can learn from "unplugging" studies or my own experience, it is that we do have a choice to opt out, but doing so can negatively affect not only ourselves but also others. Thus, we shouldn't unplug; instead, we should use whatever tools are in our possession to hack out our connected lives.

There is a part of our digital lives that we can't hack around, no matter how hard we try. As always-on participants, we trust that for-profit companies will use our data responsibly and ethically, and there is close to zero regulation over what businesses are allowed to do with our personal information. What are Mi-

---

*Microsoft shipped its Kin smartphone as a Facebook-heavy device for teens, but it quickly failed due to its high data pricing and was thus discontinued after only a few months on the market.

crosoft or Google going to do, for example, with all these photos people posted everywhere? What if your photo showed up in a series of advertisements, or what if you were caught doing something somewhat sketchy and it made its way to a blog? And who is to say that smaller, seemingly innocent private companies won't sell our information to larger groups, such as health insurance companies or marketing organizations? Already today, companies who produce software know more about us than ever before because, unbeknownst to the user, a growing number of websites and businesses are gaining extensive knowledge of user activity and psychology. In a world in which companies deliver anything-anytime-anywhere, privacy will cease to exist, whether we like it or not.

Chapter 9

iSpy:
the end
of privacy

No one expected Kris Allen to win 2009's *American Idol*—not even Allen himself. Throughout the show's eighth season, TV pundits raved about contestant Adam Lambert, who dazzled judges and audiences with his "immaculate" vocals and a show-stopping performance that even reduced Paula Abdul to tears. Allen, too, felt dubious about his own victory. "I'm sorry, I don't even know what to feel right now. This is crazy," he said on stage after hearing he had won. "It feels good, but Adam deserves this."

Allen was considered a long shot. He didn't wow judges until a semifinal round with his rendition of Kanye West's "Heartless." It was a close call: out of one hundred million votes from viewers, Allen won by less than a million votes.

After the contest's results, a company called EchoMetrix bragged that it had predicted the underdog's victory half a day before the votes were announced.[1] Making such a prediction would be difficult given the haphazard nature of *American Idol*'s voting system, in which during the show's semifinal rounds, viewers can dial in or text message their vote whenever a contestant is performing and again at the end of the episode. After the end of the show, for a two-hour time window, viewers can vote as many times as they wish. So how did this virtually unknown company foresee what the entire world seemed to be betting against?

The answer is simple: they spied on children. EchoMetrix's secret weapon was a piece of software called PULSE. "The unmatched ability to get inside privileged IM chats positions PULSE as a far more accurate predictor of the teen mindset. In fact, PULSE predicted the American Idol upset more than twelve hours before the votes were announced!" EchoMetrix said in a press release.[2]

EchoMetrix's history is peculiar. The company was previously named SearchHelp when it launched in 2004, selling a piece of software called FamilySafe that enabled parents to monitor their children's online activities. Five years later the company rebranded itself EchoMetrix and released PULSE, a tool that provided insight into today's youth to third-party marketers by aggregating data from millions of teens' instant messages, chat transcripts, blog posts, and more. With this wealth of information, EchoMetrix was able to predict the seemingly improbable winner of the eighth season of *American Idol*. The company also uses PULSE to track other marketing trends in order to determine, for example, that teens talk about iPods thirteen times more than the Zune MP3 player, and the iPhone gets four times more buzz than the BlackBerry.

EchoMetrix has since deleted its press release about PULSE's *American Idol* prophecy after attracting attention from a nonprofit group, Electronic Privacy Information Center (EPIC). EPIC put two and two together and concluded that EchoMetrix was directly violating the children's Internet privacy law by collecting data on minors without parental approval. The organization then filed a complaint to the Federal Trade Commission.[3] "EPIC's complaint describes the harm from EchoMetrix's online data collection—harm that is experienced by the millions of children and teenagers in the United States who are not aware they are being monitored," EPIC said. "The harm is also experienced by parents who unwittingly subject their children's private information to third parties for marketing purposes." In September 2010 EchoMetrix paid a $100,000 penalty to the state of New York as part of a settlement in which it agreed to not "analyze or share with third parties any private communications, information, or online activity to which they have access," including children's information, to third-party marketers.[4]

Using teenagers' private information to guess the winner of *American Idol* may seem trivial, but obviously EchoMetrix had far more information about children than it published. Think private photos, conversations about their sexual orientation, or anything children would normally jot down in a secret diary.

If a company was able to get away with monitoring and selling children's private information, imagine how much more of our private information can be exposed by our smartphones. As we move into the future, companies possessing this sort of "seeing eye" capability pose an increasingly greater threat to privacy for people of all ages in a world that's always on. The implications are even more far-reaching for the geo-aware, always-on devices that we carry everywhere, like iPhones, iPads, and GPS units.

What does this all mean for privacy? How can we be assured that our smartphones are not broadcasting our exact location wherever we go? How can we know whether smartphone applications are taking pictures of our every move or harvesting our personal data and selling it to marketing companies?

For the most part, we take a leap of faith when we believe that these for-profit companies are doing the right thing—an idea that may be naive. Consequently, to examine the state of privacy in the always-connected future, we have to look closely at cyber laws, privacy policies of major tech companies, and the tracking methods of analytics services.

Peeping Toms

While on an airplane, *Salon* journalist Simson Garfinkel noticed that his laptop was trying to connect itself to the Internet. Confused, he tried toggling his computer back into offline mode, but moments later it attempted to go online again.

He did what everyone would do in this situation: reboot. And after seeing the same behavior, Garfinkel dug a little deeper and realized that a piece of software he never installed was trying to pull an Internet connection. Even more peculiar, the hidden software was living inside Arthur's Reading Race, a children's game that Garfinkel bought to teach his daughter to read, made by the toy company Mattel.

When he called Mattel to ask about the software, a spokeswoman explained to Garfinkel that the software, called Brodcast, was only running in the background to retrieve updates or fix bugs. That may very well have been true, but the issue remained that the software was installed unbeknownst to the user and that its activities were undocumented. "While this indeed may accurately describe the company's intention in including the DSSAgent, it's pretty easy to see how such technology could cause problems," wrote Garfinkel in a *Salon* column dated June 2000. "If it wanted, the company could scan your hard drive for competing products, then flood you with offers to purchase its own similar products, or even just use that info for competitive research. Once this kind of capability is introduced, it could also be misused by a rogue employee to retrieve your financial records or credit-card numbers or to download child pornography onto your computer."[5]

Ten years later Garfinkel's prescient predictions were only beginning to unfold. In what has been dubbed a cautionary tale of how technology can impact personal privacy, high school student Blake Robbins and some of his classmates are at the center of a juicy webcam scandal in Philadelphia. In January 2010 the students filed a class-action lawsuit against Lower Merion School District, alleging that school staff were spying on students using surveillance software secretly installed on school-issued laptops.[6]

In the suit the students alleged that IT administrators were using surveillance software to spy on them in their homes by snapping webcam photos and taking screenshots of their activities. The school district denied the allegations, claiming that they installed and used the surveillance software, LANRev, only for tracking stolen or lost laptops. In response, Robbins's lawyers note that the laptop was neither missing nor stolen.

After the news broke, Lower Merion School District announced that they temporarily deactivated their webcam surveillance program. The FBI decided in August 2010 that it would not prosecute administrators because it could not find evidence that would establish beyond a reasonable doubt that anyone

involved had criminal intent. Even so, the story of Robbins un-
derscores some startling issues surrounding digital privacy. Because Internet technologies are still relatively young, and
because the government has loose regulation over broadband,
whether undisclosed remote filming of anyone, even minors, is a
federal crime under current law remains unclear.

Nonetheless, that the Philadelphia webcam scandal is stuck in
a legal gray area is strange. Whether or not Robbins's allegations
are true, Lower Merion School District has acknowledged that
the hidden surveillance software captured a substantial number
of pictures of students. This would appear to be a direct violation of the Children's Online Privacy Protection Act (COPPA),
put in effect in 2000.

The children's privacy law prohibits companies from collecting data on minors under thirteen without explicit parental consent in the form of a written letter, a fax, or an e-mail message.
When issuing the laptops, Lower Merion never informed students and their families of the tracking software in the paperwork that students signed when getting the computers,
according to Doug Young, a Pennsylvania School District spokes-
man.[7] In fact, a promotional video for the LANRev published by
news outlets online shows Michael Perbix, an IT administrator
at Lower Merion, talking about making the surveillance software
completely undetectable. Perbix says, "When you're controlling
someone's machine, you don't want them to know what you're
doing." He adds, "I've actually had some laptops we thought
were stolen which actually were still in a classroom, because they
were misplaced, and by the time we found out they were back, I
had to turn the tracking off. And I had, you know, a good twenty
snapshots of the teacher and students using the machines in the
classroom."[8]

However, even if Lower Merion was taking photos of students without their knowledge or parental consent, which seems to be a violation of COPPA, online privacy law has historically been a slippery slope, even when minors are involved. For example, back in 2000, when *Salon*'s Garfinkel confronted Mattel about Brodcast, the toy company argued that COPPA does not cover its software because the law applies to websites, not software. Even so, Mattel soon after began shipping newer Mattel products with upgraded versions of Brodcast that asked for a user's approval before installing the software.

With EchoMetrix's collection of children's personal information, the company may possibly be violating COPPA. Interestingly, in light of EPIC's complaint, the Department of Defense has prevented EchoMetrix from selling its software to military families. However, the FTC still has not replied to EPIC's complaint or taken action. Apparently, the FCC hasn't reacted to the idea that legitimate tech companies (or school districts) might use our information for any reason other than their stated purposes. In general the FCC's website reveals that the government is currently focused on cyber threats in the context of terrorists invading US networks. As a result, even though there are rules concerning user privacy, even for minors, they are not strictly enforced.

Despite the silence, EPIC is demanding that the FCC revise COPPA to use stronger wording for its rules guarding privacy in light of newer technologies, including social networking websites and location-aware smartphones. These types of innovations, EPIC argues, are poised to create even greater threats to children's privacy. "The need for the COPPA Rule has become increasingly urgent in light of new business practices and recent technological developments, such as social networking sites and mobile devices," EPIC wrote.[9]

Social networking sites and even smartphone apps do indeed create new possibilities for companies to feed off our information without our knowledge. Even an employee at Google has been accused of perverse privacy violations. In September 2010 gossip blog *Gawker* published a piece accusing David Barksdale, a twenty-seven-year-old former Google engineer, of spying on four underage teens by peeping at their Gchat logs. A source told *Gawker*'s Adrian Chen that Barksdale was virtually harassing teens of both sexes; in one incident a boy refused to tell Barksdale the name of his girlfriend, so the engineer tapped into the teen's call logs to grab his girlfriend's information and threatened to call her, said the source. In another case, when chatting with children, Barksdale was pasting quotes from a private IM behind the person's back, the source said. Barksdale also allegedly unblocked himself from a Gchat buddy list even though the teen had tried to cut communication with the engineer. When *Gawker* contacted him via e-mail, Barksdale admitted he was fired but refused to comment on why, saying only, "You must have heard some pretty wild things if you think me getting fired is newsworthy." Google's former senior vice president of engineering Bill Coughran confirmed that Barksdale was dismissed for "breaking Google's strict internal privacy policies," without elaborating on the circumstances.[10]

Our personal data is not the only thing at stake in an age that is always on. Privacy probes have even been effective against the most secretive corporation on the planet, Apple.

Data Miners

On January 26, 2010, Steve Jobs received an unpleasant surprise. Reading the newspaper, he saw that a small startup called

Flurry had traced the activity of fifty tablet devices being tested in Apple's Cupertino headquarters. This was the iPad, which Jobs was preparing to introduce to the world a day later.

Jobs realized that Flurry's analytics software was secretly embedded inside iOS applications that Apple engineers were using to test iPads. From the data, Flurry was able to assess that these devices never left the Apple campus and that many of the apps tested were content-rich books and games. Flurry deduced that these were tablet devices bolted down in Apple's test lab.

Not at all pleased, Jobs enforced a temporary blanket ban on all third-party analytics companies from inserting their software into iPhone and iPad apps. "The way they detected this was they're getting developers to put their software in their apps, and their software is sending out information about the device and about its geolocation and other things back to Flurry," Jobs said in an on-stage interview with the *Wall Street Journal*'s Walt Mossberg. "No customers ever asked about this, it's violating every rule in our privacy policy with our developers, and we went through the roof about this. So we said no, we're not going to allow this. This is violating our privacy policies, and it's pissing us off that they're publishing data about our new products."[11]

The Apple CEO says his company takes customer privacy seriously, and clearly Jobs is just as concerned about protecting his secrets as well. Any software programmer can embed analytics software to determine the best places to insert ads or to track which features are used the most frequently for targeted investments. However, because analytics software is invisible and doesn't require user permission, it should raise concerns about whether these companies are using personal data appropriately. In the case of Flurry, clearly the company crossed a line.

Apple's privacy policy suggests that third-party analytics companies can't be fully checked. Apple is a licensee of the TRUSTe privacy program, an independent nonprofit organization that reviews company practices to ensure they don't misuse customer information. However, one clause of TRUSTe's policy reveals the limits of its regulation: "TRUSTe only covers the information collected and shared by our clients. We do not monitor the uses of that information by the business partners of our clients."[12]

In other words, TRUSTe evaluates Apple's practices, but it cannot monitor third-party companies living inside Apple's platform—like Flurry. Accordingly, Apple revised its policy, effectively banning third-party analytics services, only allowing them on a case-by-case basis with Apple's written approval.

My friend David Barnard, an independent iPhone app programmer, stated that Apple's policy regarding analytics was designed, in his view, so that Apple could take back control of how developers and third parties access and use sensitive user data, doing so for the sake of protecting its own product. Barnard points out that companies like Facebook and Google trade in information—the more detailed and personalized the information, the more valuable. For that reason, he trusts Apple more than Google and Facebook: Apple is a hardware company as opposed to a data company, so its priority is the device and the experience that device provides to the customer. "If Apple didn't do this, a year from now a self-conscious woman would look down at her phone and see an ad promoting weight-loss products to overweight forty-one-year-old women with thinning blonde hair who live in a blue house and drive a black Ford Taurus," Barnard said. "And that's going to scare the crap out of her. Who's she going to blame? What product is going to be trashed in the press for enabling this kind of eerily specific advertising?"[13]

The answer to that question is, of course, the iPhone. Apple has a motivation to prevent third parties from invading customer privacy because it is a hardware company, not a data trader, and it just wants people to keep buying iPhones, Barnard reasons.

Barnard's lack of trust in Facebook isn't unwarranted. Although Facebook, too, is a TRUSTe licensee, this badge of honor didn't prevent the company from major public scrutiny regarding privacy. Throughout much of 2010 critics slammed Facebook for inundating its website with confusing options that potentially manipulated users into sharing more information than they truly had opted for. In mid-2010 EPIC filed a thirty-eight-page complaint against Facebook with the FTC, demanding that Facebook remove new features introduced in mid-April that forced users to share some personal information.[14] EPIC was referring to Facebook Connect, which allows users to log into other websites using their Facebook credentials. In one section of Facebook's policies the site admits that when a user accesses a Facebook application or a website using Facebook Connect, the software shares that user's name, profile pictures, gender, user IDs, connections, and friends. The user has no ability to opt out of sharing this information with these sites and apps, and what Facebook is doing with this information is unclear.

"Facebook now discloses personal information to third parties that Facebook users previously did not make available," EPIC said in its complaint. "These changes violate user expectations, diminish user privacy, and contradict Facebook's own representations." Furthermore, the website's increasingly complex privacy policy gets longer every year; by 2010 it was 5,830 words—even longer than the US Constitution.[15]

In response to the backlash, Facebook's CEO Mark Zuckerberg admitted in a *Washington Post* guest column that Facebook "missed the mark"—without explicitly apologizing, of course—and said the site would issue a major update that streamlined privacy settings.[16] Even with the updated privacy settings, privacy issues still remain a major topic of criticism for Facebook.

As for Google, the corporation didn't even qualify as a TRUSTe member. "Google is not meeting a basic Internet best practice for websites that care about consumer trust," TRUSTe said in a public blog post.[17] The nonprofit company explained that privacy experts have agreed on the best practices for privacy, and Google falls short of meeting the criteria. For example, one important TRUSTe requirement is that company websites include a link to their privacy policy on their home page. Google doesn't comply but does state in its privacy policy, with very vague wording, that it provides some of its users' personal information to third-party advertisers: "In some cases, we may process personal information on behalf of and according to the instructions of a third party, such as our advertising partners." Therefore, the eerily targeted ad that Barnard imagined about a forty-one-year-old woman driving a black Ford Taurus wouldn't be too far off from what we are likely to see in a phone Google produces or a mobile website Google builds.

Indeed, perhaps to no one's surprise, Google came under massive scrutiny in 2010 when the company admitted that its Street View cars—vans that drive around streets collecting imagery for Google Maps—were mistakenly collecting samples of data sent from non-password-protected wireless networks.

This was no small mistake when carried out by a company as huge as Google. The Street View cars collected six hundred gigabytes of payload data over thirty countries, the search giant admitted. "We screwed up," said Sergey Brin, Google's cofounder, during the company's I/O developer conference in May 2010. The company promised to delete the payload data it erroneously collected.

Even as groups are working to get the government to clarify and strengthen regulations for these kinds of violations of privacy, always-on technology rapidly outpaces preexisting legislation. Furthermore, by the time any new laws emerge, new technologies have already created more possibilities for privacy to be violated. This issue is already apparent in cell phone usage.

The Right to Remain Searchable

Imagine that a twenty-two-year-old Oakland man named Phil gets pulled over for running a stop sign. A police officer approaches the passenger window and determines that Phil looks suspicious and decides to arrest him—because running a stop sign is an arrestable offense. While patting down Phil, the officer finds a pack of cigarettes and an iPhone in his pockets. Under the search-to-arrest doctrine, the officer is entitled to open the pack of cigarettes and search it without a warrant or even a probable cause to believe there is anything illegal inside. Further, under the same doctrine, the officer also has a right to a warrantless search of Phil's iPhone.

The iPhone (and any smartphone, for that matter) can be considered the digital equivalent of a closed container, which police officers can search thoroughly during an arrest, according to

Adam Gershowitz, an associate professor at the University of Houston Law Center. That's because even though society and technology have transformed dramatically over the past few decades, the Fourth Amendment, which guards citizens against unreasonable searches and seizures, has remained static.

This is not a new observation. For years courts have accepted digital information as evidence; they see no conceptual difference between physical containers and gadgets containing data. Before the iPhone, police officers used information retrieved from pagers and conventional cell phones as admissible evidence against criminals. However, with media-rich, all-in-one portables such as the iPhone, the situation changes tremendously. Simply by searching an iPhone, police officers can rightfully gain access to a treasure trove of personal information. In addition to text messages, contacts, and call histories, an iPhone holds far more pictures than could be stored on a conventional cell phone and displays them in much clearer detail. Furthermore, the data contained inside third-party apps can potentially tell a person's life story.

The iPhone may certainly come in handy for nabbing criminals, but the implications of lost privacy are relevant to hundreds of millions of people carrying smartphones today. If Mel Gibson were pulled over and arrested, a police officer could potentially read his text messages and e-mails and listen to his voicemails to learn quickly that he is an anti-Semitic misogynist. An arrestee's iPhone Facebook app could reveal that he's having an affair with a coworker in his office, or a politician's data could reveal that he's a homosexual—a secret he has been trying to keep from the public. These pieces of information, as they are not crimes, would most likely be irrelevant to any criminal offense, so they wouldn't be brought up in court, but there is always a chance that these secrets can get released to be public. "In the search for incriminating information, officers will no doubt come into contact with extremely sensitive personal information that is not remotely illegal but which is nevertheless highly embarrassing," Gershowitz says. "And while such embarrassing, but not incriminating, information probably would not be admissible in a prosecution, its discovery would cause emotional distress. Moreover . . . it sometimes manages to find its way into the public domain."

Ever since the release of the iPhone in 2007, law professors have been debating about potential solutions for addressing new technologies in regard to the Fourth Amendment. One possibility is to limit the number of steps a police officer can take when searching a phone—for example, five actions such as 1) powering on the phone, 2) launching the Internet browser, 3) opening the browsing history, 4) opening a text message, and 5) opening the second most recent text message. Another possibility is to limit searches to only those that are related to an arresting or suspected offense. For instance, if a person were suspected of being a drug dealer, the officer would have the right to launch the text message app because drug transactions are often carried out through texts. Both these options rely on the officer's honesty, of course, though the same can already be said about how officers handle searching physical property.

Some police officers are already taking advantage of extracting data from smartphones. My friend Jonathan Zdziarski, a popular security researcher, teaches digital forensics classes specifically on the iPhone to law enforcement. In his seminar Zdziarski instructs officers on hacking the iPhone to crack pass codes, bypass encryption, and recover potential evidence that a suspect may have tried to hide or destroy.

Many of these methods rely on exploiting the iPhone's security weaknesses. For example, one peculiarity of the iPhone is that it takes a screenshot whenever you switch to a new application. The image of a user's last action is stored so that it can be restored when she returns to it. The device also creates this image to produce a shrinking effect when an app is closed. With some clever hacking tricks, Zdziarski can successfully retrieve all these secretly stored images in order to view messages that a suspect tried to erase, websites he had viewed, and so on. "I'm kind of divided on it," Zdziarski told me when he discovered this in 2008. "I hope Apple fixes it because it's a significant privacy leak, but at the same time it's been useful for investigating criminals."[18] As of this writing, however, Apple still has not patched this security hole. In the meantime some of the iPhone's security flaws have been used to gather evidence against criminals convicted of rape, murder, or drug deals, according to Zdziarski. And the only reason we haven't heard about these stories is that no court has been called on to address the constitutionality of searching an iPhone.

Currently, nothing is changing about the Fourth Amendment to guard against digital invasions of privacy. In this sense, having "your life in your pocket" with an iPhone can be both a curse and a blessing. With barely any regulation over websites, smartphone apps, and analytics companies as well as a frozen Fourth Amendment, the iPhone can potentially give anyone access to anything about a person at anytime, anywhere that person may be.

Does Privacy Matter?

A philosophical divide arises over whether or not privacy is important. One school of thought argues that privacy doesn't even matter anymore. With the rise of social networking websites and smartphone apps, people are willingly giving away all sorts of personal information everywhere they go. For example, FourSquare, a site that enables people to share their location by "checking in" to entertainment venues, restaurants, and other attractions, has collected over 750 million check-ins. Twitter, a microblogging service designed to immediately disseminate information into the wide-open public, is one of the fastest-growing platforms ever, with 300 million users. Finally, Facebook, despite its constant backlash for poor handling of privacy, has accumulated well over 845 million users who share more than 30 billion pieces of content each month. The widespread enthusiasm for always-on interpersonal connections is indisputable proof that people are willing to sacrifice privacy in exchange for services richly tailored to their unique lives.

Perhaps this wealth of personalized services that we receive in exchange for privacy is the "modern trade." This is what *Wall Street Journal* writer Jim Harper argued in an August 2010 column. He argues that the reason we receive such rich, useful data is because we trade some of our personal information. Harper sums it up succinctly:

> The reason why a company like Google can spend millions and millions of dollars on free services like its search engine, Gmail, mapping tools, Google Groups and more is because of online advertising that trades in personal information.
>
> And it is not just Google. Facebook, Yahoo, MSN and thousands of blogs, news sites, and comment boards use advertising to support what they do. And personalized advertising is more valuable than advertising aimed at just anyone. Marketers will pay more to reach you if you are likely to use their products or services. (Perhaps

online tracking makes everyone special!) If Web users supply less information to the Web, the Web will supply less information to them. Free content won't go away if consumers decline to allow personalization, but there will be less of it.[19]

Regardless of one's stance on privacy, there is a haunting side effect to being always on: data lives forever, for better or for worse. Will Moffett, an independent web developer, launched a satiric website mocking Facebook's privacy flaws called YourOpenBook.org. When visiting YourOpenBook.org, you can type in a keyword and hit "Search," and YourOpenBook.org shows a listing of every public Facebook status update containing that term. Simply by typing in a few searches and browsing through the postings, one can immediately see that many users share more than they may be aware of: searches such as "I cheated on my husband" reveal people openly admitting adultery; a search string for the word "blowjob" sometimes shows people advertising special services meant for their friends; and a search for racial slurs sheds disturbing insight on some people's unbridled prejudices.

Moffett said he designed YourOpenBook not only to make fun of Facebook but also to share the message that even if privacy doesn't truly exist in the digital age, people still have a right to control what's left of their personal lives online. He believes people still need privacy. "When we make mistakes, we learn from them in a private sector with our friends and our families," he told me during a conference. "Without privacy, when we make mistakes they'll haunt us forever. How can we learn and move on?"[20]

Perhaps we have already given up our digital privacy, but we still have control over boundaries. *Merriam Webster*'s definition of privacy is "a place of seclusion." In a world that's always on, where seeing eyes can peek through software as easily as dust can travel through a window screen, we are plugging into someone else's place (i.e., a company's online server), and when we do, seclusion is lost. In a modern online context a violation of privacy may only occur when we are manipulated into sharing more than we were told we would be sharing. Online privacy advocates criticize online services when they are unclear or dishonest about what they are doing with our data, not when they are using our data—because, of course, they are.

Perhaps in regard to our privacy online, we should think like the Amish—that is, we should ask, How can we use these technologies so that data can benefit us rather than cause harm? And technology *can* harm. In September 2010 a Rutgers University freshman, Tyler Clementi, was kissing another young man in his dorm room while his roommate Dharun Ravi was secretly videotaping him on a webcam. "Roommate asked for the room till midnight," Ravi wrote on Twitter that night. "I went into Molly's room and turned on my webcam. I saw him making out with a dude. Yay." According to authorities, Ravi then streamed the video on the web to friends. Seventy-two hours later Clementi updated his Facebook wall with a chilling message: "Jumping off the gw bridge sorry." And he leapt to his death.

In this constant, rapid exchange of information we engage in online, some of us forget that social boundaries still exist. This is understandable. After all, we are being brought up to share, share, share; where we draw the line can get awfully confusing. I presume Ravi thought of his stunt as a practical joke, not a malicious act, but we should take his mistake as a lesson for ourselves: for the most part, we are subject to harming someone online when we share information we probably should not publicly share to begin with, regardless of whether we are sharing it with friends or with strangers. We should think before we tweet as often as we should think before we speak.

Chapter 10

perfect vision

When Alex Dejong was forty-four years old, doctors gave him a grim prognosis: he had developed a brain tumor that was crushing his optic nerve, causing him to go blind. Dejong was devastated. His loss of vision threatened to destroy his longtime career as a professional photographer, taking away his gift for seeing the world in a different light.

Fortunately, the blindness sunk in gradually, so Dejong had time to think and adapt to the handicap. First, he lost his peripheral sight, maintaining only his central vision. Then, finally, after a few operations to remove the tumor, everything went dim and all Dejong could perceive was light or dark coming from certain directions.

Religiously trained in photography, however, Dejong wasn't ready to give up his craft just yet. He turned to gadgets. Carrying around a Nokia N82 cell phone, Dejong used assistive software to translate sounds into visuals in his mind. The software, called vOICe, analyzes the light the handset's camera detects and plays different sounds depending on the brightness, thereby helping the blind make pictures out of sounds. After stitching together a mental image of his surroundings, Dejong snapped photos with his Canon and Leica digital cameras.

Still, the problem remained that Dejong could not see his own photos to review or edit them, so he hired an assistant. Later, the third-generation iPhone came along to help Dejong hear what he couldn't see. The iPhone included built-in software called VoiceOver, which read back anything a user places his finger over on the screen—e-mails, web pages, system preferences, and so on. Dejong downloaded two photo-editing applications for the iPhone called CameraBag and TiltShift, which perform automated editing tasks he couldn't otherwise do, and with the help of VoiceOver, he was able to process and then upload his photos to a website.

So just like that—with a Nokia helping him "see" and an iPhone enabling him to edit and post his photos on the web—Dejong was a full-fledged photographer again. "With the iPhone and a lot of the photography apps that a lot of people are using, I have my entire workflow, and I can do it in five minutes," Dejong told me. "In this way, the iPhone is a remarkable gift. It has really opened up my world."[1]

Thus, Dejong used a combination of software and hardware to enhance his own mental faculties and overcome a handicap. In other words, he used data to extend his perception of the real world. This is a very early example of what technologists call "augmented reality," a future in which data becomes seamlessly interlaced with the physical world in order to enhance our everyday perception, thereby giving us the eyes of a cyborg. Augmented reality is the inevitable next step for always-on, mobile technologies that deliver to us anything-anytime-anywhere, and it is a vision that is only beginning to unfold.

Augmented Reality

As you shove your way through a crowd in a baseball stadium, the lenses of your digital glasses display the names, hometowns, and favorite hobbies of the strangers surrounding you. You claim a seat and fix your attention on the batter, and his player statistics pop up in a transparent box scrolling in the corner of your field of vision. A food vendor carrying a tray of snacks waddles around the bleachers, and as you gaze at the cotton candy, hot dogs, and bags of buttery popcorn, small digits pop up in front of each item showing their price.

Although not possible today, the emergence of more powerful, media-centric cell phones is accelerating humanity toward this vision of augmented reality, in which data from the network overlays our view of the real world. Already software developers are creating augmented reality applications and games for a variety of smartphones, making a phone's screen display the real world superimposed with additional information such as the location of subway entrances, the price of houses, or Twitter messages nearby people are posting. And naturally, publishers, moviemakers, and toymakers are pouncing on early versions of this technology to make their advertisements more in-your-face than ever. "Augmented reality is the ultimate interface to a computer because our lives are becoming more mobile," said Tobias Höllerer, an associate professor of computer science at UC Santa Barbara, who is leading the university's augmented reality program. "We're getting more and more away from a desktop, but the information the computer possesses is applicable in the physical world."[2]

Technologists have been tinkering with augmented reality for decades. Tom Caudell, a researcher at aircraft manufacturer Boeing, coined the term "augmented reality" in 1990,[3] using it to describe a head-mounted digital display that guided workers through assembling electrical wires in aircraft. The early definition of augmented reality was an intersection between virtual and physical reality, in which digital visuals are blended with the real world so as to enhance our perceptions.[4]

With smartphones, we've already made significant progress blending data into our everyday lives, and we're not stopping there. Futurists and computer scientists continue to raise their standards for a perfectly augmented world. Höllerer dreams that augmented reality will reach a state in which it does not rely on a predownloaded model to generate information. That is, he wants to be able to point a phone at a city that is completely unfamiliar, download the surroundings in real time, and then output information on the fly. He and his peers at UCSB call this idea "Anywhere Augmentation."

But we have a long way to go—perhaps several years—before achieving Anywhere Augmentation, Höllerer told me. Limitations in software and hardware stifle the realization of augmented reality. For hardware, cell phones would require superb battery life, computational power, cameras, and tracking sensors; for software, it requires much more sophisticated artificial intelligence and 3-D modeling applications. Above all, this technology must become affordable to consumers. The best possible technology available today would cost nearly $100,000 to produce a solid anywhere augmented–reality device.

Regardless, a large number of augmented reality enthusiasts

are already getting started. For basic consumer needs, software developers are focusing on offering smartphone apps that help us enhance our perception of the real world. For example, Layar, a company based in Amsterdam, made an augmented reality browser for Android smartphones and iPhones. The Layar browser looks at an environment through the phone's camera, and the app then displays houses for sale, restaurants, shops, and tourist attractions. The software relies on downloading "layers" of data provided by third-party developers coding for the platform. Thus, although the information appears to display in real time, it is not truly real time: the app can't analyze data it hasn't downloaded ahead of time, so in a sense it is *pre*-augmented reality with plenty of gaping holes—uncharted pieces of our world.

Augmented reality isn't going to stop just at visuals either, according to Raimo van der Klein, CEO of Layar. He expects applications to move beyond augmenting vision and expand to other parts of the body. "Imagine audio cues through an earpiece or sneakers vibrating wherever your friends are," van der Klein told me.[5]

Augmented reality gets a little creepier with an app for Google Android phones called Recognizr, a piece of software designed to identify people simply by snapping a photo of them. In development by Swedish company Astonishing Tribe, the Recognizr app uses recognition software to create a 3-D model of a person's mug. It then transmits that model to a server and matches it with an image stored in the database. The online server performs facial recognition, shoots back a name of the subject, and then links the user to the subject's social networking profiles. In its current state, however, Recognizr can only identify a small number of profiles stored inside its database. Furthermore, because smartphone chips and broadband networks need to get speedier and more powerful before they can crawl billions of photos across the web to identify and recognize any random person, several years will probably pass before a fully working product hits the consumer market. Still, this is a somewhat disturbing example of what our data-injected future has in store for us.[6]

Some researchers are focusing on using augmented reality to enhance one of the most lucrative industries in tech: gaming. Georgia Tech's Augmented Environments Lab, for example, has made an augmented reality zombie shooter called ARhrrrr. The game involves pointing a phone's camera at a map containing markers, and a 3-D hologram of a town overrun by zombies appears on the phone's screen. Using the phone, a user can shoot the zombies from the perspective of a helicopter pilot. What's more, one can even place (real) Skittles on the physical map and virtually shoot them to set off virtual bombs.[7]

Marketing companies, of course, would never pass up the opportunity to make advertising more attention grabbing than ever by popping promotions in front of us. For example, augmented reality was used to promote the sci-fi blockbuster *District 9*. There was a "training simulator" game on the movie's official website; however, before a user could access it, he had to print a postcard containing the *District 9* logo and hold it in front of a webcam. The postcard contains a marker, and once the game detected that marker in the webcam video, it overlaid a 3-D hologram of a *District 9* character on the computer screen. From there, players could click buttons to fire a gun, jump up and down, or throw a human against a wall in the game.[8]

However, augmented reality isn't truly useful in a static desktop environment, Höllerer said, because people's day-to-day realities involve more than sitting around all day (outside of work, at least). This is why smartphones, which include GPS hardware and cameras, are crucial to driving the evolution of augmented reality.

Brian Selzer, cofounder of Ogmento, a company that creates augmented reality products for games and marketing, recognizes the need for augmented reality to go mobile. He said his company is working on several projects coming in the near future that will help market mainstream movies with augmented reality smartphone apps. For example, movie posters will trigger interactive experiences on an iPhone, such as a trailer or even a virtual treasure hunt to promote the film. "The smartphone is bringing AR into the masses right now," Selzer said. "Every blockbuster movie is going to have a mobile AR campaign tied to it."[9]

"We're doing as much as we can with the current technology," Selzer added, discussing the overall augmented reality developer community. "This industry is just getting started, and as processing speeds speed up, and as more creative individuals get involved, our belief is this is going to become a platform that becomes massively adopted and immersed in the next few years."

In an always-on mobile future, the phenomenon of being able to have anything-anytime-anywhere will culminate into an ideal augmented reality. Once exponentially more powerful mobile technology leaps from our smartphone and into wearable apparel, data will inevitably enhance our perceptions of everyday reality, and this will change everything, including how we work and how we interact.

Wearable Data

To learn the secrets of human history, archaeologists dig into our earth and study each and every layer. By its very nature, this process is destructive and physically unreconstructable, so archaeologists record their digs with diagrams, notes, and 3-D models—a difficult task, as each team member is usually working on one particular section of a site.

Steve Feiner, a computer science professor at Columbia University, pioneered some of the earliest augmented reality technology dating back to the 1990s.[10] One of his recent projects was called Visual Interaction Tool for Archaeology (VITA), a system that consisted of a bulky head-worn display to visualize 3-D terrain data and embedded multimedia, and this interacted with a tracked handheld display, another high-resolution display, and a multitouch projected table surface. All these devices worked together to enable archaeologists to take a virtual tour of the entire dig site: one mode enabled users to see individual layers at specific parts of a site, and another "world-in-miniature mode" presented a small-scale virtual model of the entire dig site in order to see the spatial relationships between different layers. In short, VITA allows archaeologists to collaboratively discuss and analyze a digital reconstruction of a dig site so as to tie together their physical analyses.

Various industries could benefit from a gadget similar to the VITA. Architects could superimpose 3-D models onto their vision while inside a construction site in order to devise more innovative structures than ever before. Repairmen faced with challenging tasks, such as plugging BP's oil spill, could calculate potential fixes while observing the scene of disrepair. Mechanics donning headwear could know precisely how to tune up each and every vehicle by displaying 3-D models of the car's insides accompanied with step-by-step instructions. The list can go on and on.

However, developed in 2004, VITA was a prototype device with cumbersome hardware and limited marketability, so the chances that it will become commercially available anytime soon are slim. Echoing some of Höllerer's concerns about smartphone weaknesses, Feiner highlighted challenges that today's mobile devices must overcome before achieving commercially viable augmented reality. Above all, tracking capabilities in today's GPS-embedded smartphones are still too imprecise: depending on the time of day and where it is, even with sensors such as gyroscopes, magnetometers, and accelerometers, a smartphone's tracking sensors can still be off by tens of meters when "pinpointing" its geographical location. Also, a camera's field of view tends to be too large compared to what can be displayed on the smartphone's small screen (e.g., the iPhone's display is 3.5 inches). For this reason, in today's smartphone augmented reality apps (such as the Yelp app for iPhone), when a user points her camera at a location, the phone can be extremely inaccurate when overlaying corresponding data. Because of these shortcomings, augmented reality apps today are mostly a novelty and hardly useful.

Moreover, before it breaks into the mainstream, augmented reality needs to become fashionable. The current state-of-the-art eyewear is ugly, bulky, and uncomfortable. Even with smartphones, there is still something socially unacceptable about walking around while staring at the world through a phone. "Whenever I pull out phones to demonstrate [augmented reality], my wife steps back a couple of paces to indicate she's not with the crazy guy holding his phone and walking around," Feiner said.

With all that said, Feiner is still optimistic that 2011 marks the beginning of the decade when useful and practical augmented reality will become usable to the mainstream. He believes this because smartphones are already much smaller, lighter, cheaper, and more powerful than most of the technology Feiner was tinkering with to create augmented reality applications in 1996. At the pace that they are advancing technologically, augmented reality may inevitably break into the mass market within the next ten years.

Feiner foresees that the first very practical consumer application for augmented reality is lightweight, wearable computing that displays simple data: text. For example, a pair of augmented reality glasses could act as an invisible teleprompter that one sees at all times. Such a device could enable professionals to give a slideshow presentation without breaking eye contact, help academics discuss research without having to flip back and forth between sheets of paper, and provide journalists with their notes displayed in their field of vision as they interview people. Overall, an anything-anytime-anywhere teleprompter has great potential to advance our ability to engage with the real world.

With augmented reality applications, our conventional definitions of intelligence and intellect are clearly evolving as a result of technology that delivers anything-anytime-anywhere. Once augmented reality is optimized, having 20/20 vision will no longer be sufficient: if we're not seeing data, we're not seeing.

The Living Cyborg

For almost twenty years professor Thad Starner has worn a computer screen over his left eye.[11] The 640-by-480-pixel eye display is hooked up to a Wi-Fi-powered ultramobile PC that is stuffed into a bag slung around his shoulder. Starner also carries a smartphone in his pocket to get a cellular connection whenever he wants it, as well as a one-handed keyboard called a Twiddler, which enables him to type on average seventy words per minute.

It is fair to say that Starner is one of the few enthusiasts today who is truly always on. He lives and breathes data.

In an interview with Gartner Research, Starner said he realized this was a lifestyle he desired back when he was an undergraduate at MIT, when he was spending $20,000 a year on tuition and not remembering any of what he was learning in class. He could jot down notes during a lecture, but that would detract from his attention to the professor. So, for him, embedding a note-taking system into an eye display was a viable solution. Beyond academic purposes, Starner uses his heads-up display everywhere he goes. When chatting with a friend or a colleague, he reads notes projected in front of his eye that contain every previous conversation with that person so he can pick up exactly where they left off. During faculty meetings at Georgia Tech, other faculty members will often ask Starner to send notes or e-mails of something they just decided to do.

In some ways, the Internet has blended into Starner's everyday behavior as well. He learned to begin constructing his sentences while quickly searching for the proper information to fill in at the end. So, for example, if someone asked Starner, "Who was the first son of Henry III?" Starner could reply, "I believe it's . . ." while Google searched the answer, and then finish with "Edward I of England."

"Google on the eyeballs is really powerful," he said. "I use the wearable computer to augment my own thought, my own intellectual capabilities, as opposed to just my communication abilities."

But why notes, and why not video? Starner pointed out that because one can't simply type a keyword or utter a phrase to load a specific frame, video is a pain to search, plus it takes up copious amounts of memory. And even though he's constantly juggling around notes, Starner said his memory has actually gotten sharper. He attributes this to how he records his notes while engaging with the environment. Normally, when we have a conversation, we hear a person talk—and that's it. However, the human memory works on repetition, so when Starner's having a conversation, he's listening, typing it down as a note, and still seeing the note as he types more notes—three repetitions.

Though Starner's wearable computer delivers many benefits, there are some detriments as well. Of course, he is known as "that guy" who wears a computer in front of his face. This has also led to some awkward social encounters. For instance, one time a stranger on the street asked Starner for the time, and he immediately turned, said, "3:53," and continued walking. However, "The person got really upset," Starner explained, "caught up with me and said angrily, 'How do you know?' They thought that I was blowing them off, that I made up the time since I didn't look at my wrist. I said, 'This is a computer display and the time is right in front of my eyeball.'"

Nonetheless, Starner thinks that, as people get used to being always on with devices such as iPhones and BlackBerries, it is only a matter of time before wearable computing becomes more affordable and socially acceptable. He dreams of a future when we have ESP-like capabilities thanks to data flowing seamlessly into our thought streams.

Starner's wearable computer offers a glimpse into a potential future for augmented reality as it also raises a number of questions. What does it mean to "know" something if data is immediately accessible right in front of your eyes? With the example of the Layar app, a person can potentially visit a foreign country for the very first time in his life, point his phone's camera at a restaurant, and "know" whether regulars think it is a good place to eat.

Going forward, imagine how a matured version of this application could enhance how we view everything: point your phone at a car to immediately identify the year it was made, or scan a famous monument to learn some key facts about its history. Consequently, the need to memorize is already diminishing in importance so long as we carry a general-purpose device.

Another major question that arises is, What is intelligence? We traditionally define a person's intelligence by his or her ability to quickly and accurately comprehend information. In Starner's case, however, his always-on eyewear enabled him to perform an Internet search as quickly as he could think. Thus, with the assistance of data, anyone can have an edge when solving a problem or learning new information. Everyone has access to a wealth of information, so having knowledge no longer seems all that advantageous. As a result, our conventional view of intelligence is poised to morph as well.

Perhaps in a world that is always on, an intelligent person will be defined as a curator of information rather someone who has accumulated a weight of knowledge. As such, the top performers will likely be those who are best at distinguishing high-quality and accurate data versus junk. Although *The Shallows* author Nicholas Carr and many web cynics wonder if the Internet is making us dumber, perhaps a different way to look at this question is that the definition of intelligence is changing. Even though our brains may in fact become rewired, with the help of data everywhere, we are also becoming more capable than we have ever been before. "I don't know about people getting smarter, but I think there's just going to be amazing things in terms of increasing your social IQ," Feiner said.

Finally, another inevitable question arises: What do we give away in exchange for this highly personalized type of technology?

The obvious answer is personal information, and in order to bring augmented reality into fruition, we will have to trade data much more revealing than words and numbers; we will have to share pictures everywhere we go.

Eyes Everywhere

During the 2009 presidential inauguration, CNN.com asked the public to snap photos of the moment Barack Obama was taking the oath and submit them to its website. CNN collected six hundred photos snapped from many different angles and distances and then stitched them together into a richly navigable three-dimensional model of the event.[12] Zoom in and you can see Obama accepting the oath; zoom out and you get a long shot of the entire crowd. Rotate the angle and you can look around the US state capitol as if you were there turning your head.

To compile the 3-D panorama, CNN used Photosynth, a collaborative project between the University of Washington and Microsoft. Photosynth analyzes multiple photos taken of the same area, detecting the features in each photograph to see which ones contain the same objects—for example, the side of a microphone or the front of a podium. Then, by analyzing the angles and distances of different features, the software can generate point clouds for each object to determine where to position them in relation to one another. Eventually the software stitches the photos together into a 3-D panorama.

Columbia University's Feiner believes Photosynth, or a technology similar to it, will fix the location-tracking weaknesses in today's smartphones to help augmented reality break into the mainstream. The idea is that if we accumulate a massive number of pictures of the same thing from myriad people, we can stitch them into models of different locations. Each new image a phone uploads can contain the GPS coordinates of where the photo was snapped, and once there is a large enough quantity of photos taken of a place, an augmented reality application can potentially triangulate a very precise measurement of a user's location. All a person would have to do to pinpoint her location is snap a photo.

In addition to Microsoft, Google is strongly positioned to offer a precise location service using photos as well. The company already hosts Google Earth and Google Maps, featuring satellite imagery and street photos shot with Google's Camera Car. On top of that, the search giant is lab testing Google Goggles, which promises to allow users the ability to search by shooting photos. Snap a photo of a restaurant, a book, a DVD, and so on and a user can get Google search results of the captured image. In doing so, we are sharing photos with Google, and if those images are stored in a database, Google could also triangulate photos containing GPS coordinates to determine a smartphone's exact location.

A photo-based location service would take an extraordinary amount of time to optimize, but it is a viable solution for the weak tracking capabilities in today's smartphones. In exchange we get applications that can beam us data based precisely on where we are, thus unlocking some of the dreams of "anywhere augmented reality." However, the obvious trade-off involves people taking pictures all the time, everywhere they go—a digital model of our world in real time, or a society of seeing eyes. This will likely disturb a number of people concerned about privacy. Then again, this isn't too far off from what people are already doing with social networking websites such as Yelp, Facebook, FourSquare, and Twitter, Feiner said. Hundreds of millions of social networking users are already publicly documenting information about where they are, what they are doing, or where they are eating. "We're going back to the age of the little European town where there was some grandma sitting out, leaning over the terrace and watching the street, and if anything funny happened she'd be calling the police," Feiner said. "There was a time when people liked that. It just may be that people will get used to it."

With millions of people toting around smartphones, we already seem to be getting used to the modern trade of giving up privacy, control, and some freedom in exchange for anything-anytime-anywhere. This is simply what we have to do when we are always on.

afterword

On the morning of October 4, 2011, I was in the audience at San Francisco's Moscone Center, where Tim Cook, Apple's then-interim CEO, was holding a press conference to introduce the fifth version of the iPhone. There was something off about the event—perhaps a lack of genuine excitement, the kind you'd normally see when Apple executives uncorked a secret that they had kept bottled up for a while. Of course, the conference was lacking the charismatic presence and dynamic gusto of Steve Jobs, who was on medical leave at the time. Cook, Apple's new gray-haired, placid leader, just wasn't quite the same. But other than that, there was still something odd about the mood that I couldn't quite put my finger on.

It all made sense when Jobs died one day later. I was moved when I realized he had held on just long enough to see his company introduce "one more thing"—and to know he had passed the torch to good hands. Just a few months after his death, it would seem evident that he did. Riding record holiday sales of the iPhone and the iPad, Apple climbed to become the most valuable company in the world, surpassing Exxon Mobil, the energy company. Not bad for a computer maker that was close to going broke fifteen years earlier. In the year after his death, the company continued turning in record results to become the world's No. 1 smartphone maker, and its iPad remained uncontested in the swelling tablet market. Apple's near-term numbers, however, only illustrate one thing: Apple after Steve Jobs—not Apple under the leadership of Tim Cook. Apple executives have long asserted that the company plans its products two years in advance, and therefore we won't see an iPhone or Mac purely under the direction of Cook for some time to come.

That's not to say Jobs's importance to Apple was exaggerated. The reason his company continues to do so well is because he planned far in advance. Before his second medical leave in 2011, Jobs hired Joel Podolny, dean of the Yale School of Management, to lead a program called Apple University. Podolny hired a team of business professors to study Apple's recent history and write a series of case studies about the company's most significant decisions. The purpose of this critical self-analysis was to ensure that Apple would remain Apple in the event that Jobs were to depart, avoiding the mistakes the company made in the past, and continuing to optimize its most successful strategies.

That said, a year after Jobs's death, some of Apple's products have noticeably lost polish. The iTunes software is the most prominent example, and it reflects just how big Apple's operation has become. What was once just a music player and music store is now also a massive emporium for software, books, movies, and TV shows. On the iPhone 4S, these sub-stores load slowly and have so much content that they're overwhelming to navigate. Other features, such as backing up information on an iPhone, don't work smoothly. Other than a clunky iTunes, the iPhone 4S's most famous feature called Siri, a personal assistant that a user can control with voice commands, is widely regarded as one of the most unreliable products Apple has ever released, tainting the brand name of its famous phone. But is this degradation in performance due to the lack of the Jobs, or is simply the natural result of a company that has gotten so big that it can no longer be as tight as it once was?

Some employees at Apple note a subtle change in the company's culture. Cook, they say, is more communicative than his predecessor. The overall result, they say, is an Apple with a bit more discipline and predictability, whereas before, the temperamental Jobs was the type who would make decisions with his gut, sometimes rash, sometimes brilliant. Beyond that, however, most Apple employees I've spoken with say not much has changed at all.

In an on-stage interview, Apple's new CEO said that he wasn't going to try to be the same as Jobs, but he also wasn't going to try to be very different. And perhaps that's the best tipoff for the direction of a post-Jobs Apple: it's unlikely Cook will radically change the core strategies that made the company so successful. Why stop here?

notes

Prologue

1. Brian Chen, "Man Buried in Haiti Rubble Uses iPhone to Treat Wounds, Survive," *Wired,* January 20, 2010, http://www.wired.com/gadgetlab/2010/01/haiti-survivor-iphone/.

2. Adam M. Gershowitz, "The iPhone Meets the Fourth Amendment," January 15, 2008, http://ssrn.com/abstract=1084503.

3. Chuck Squatriglia, "Ford Brings Smartphone Apps to Your Dashboard," *Wired,* April 20, 2010, http://www.wired.com/autopia/2010/04/ford-sync-applink.

Chapter I

1. Macworld Expo conference in San Francisco, January 9, 2007, attended by author.

2. John Gruber, "The iPhone's Funny Price," DaringFireball, May 1, 2007, http://daringfireball.net/2007/05/iphones_funny_price.

3. David Meyer, "Windows Mobile Loses a Third of Market Share," ZD Net UK, November 13, 2009, http://www.zdnet.co.uk/news/networking/2009/11/13/windows-mobile-loses-nearly-a-third-of-market-share-39877964/.

4. Charlie Sorrel, "Nokia Kills Symbian, Teams Up with Microsoft for Windows Phone 7." Wired.com, February 11, 2011, http://www.wired.com/gadgetlab/2011/02/microsoft-and-nokia-team-up-to-build-windows-phones/.

5. Greg Kumparek, "Apple Sold 14.1 million iPhones Last Quarter, Over 70 Million since Launch," MobileCrunch, October 18, 2010, http://www.mobilecrunch.com/2010/10/18/apple-sold-14-1-million -iphones-last-quarter-over-70-million-since-launch/.

6. Michael Arrington, "The iPhone App Store Has Launched," TechCrunch, July 10, 2008, http://techcrunch.com/2008/07/10/app -store-launches-upgrade-itunes-now/.

7. Horace Dediu, "Apple Has Accepted Nearly 400,000 Apps in 2.5 Years," asymco, December 27, 2010, http://www.asymco.com/ 2010/12/27/apple-has-accepted-nearly-400000-apps-in-2-5-years/.

8. "Computers and Information Systems," *Encyclopædia Britannica* Online, http://www.search.eb.com/eb/article-91786.

9. Gillian Wee, "Netscape to Be Shut Down After Failing to Win Users," December 28, 2007, http://www.bloomberg.com/apps/ news?pid=newsarchive&sid=a.QrjKUdHrrQ&refer=us.

10. "Netscape Communications Corp.," *Encyclopædia Britannica* Online, August 19, 2010, http://www.search.eb.com/eb/article-92983.

11. Joe Hewitt, interview with author, July 14, 2010.

12. Matthew Gray, "Measuring the Growth of the Web: June 1993 to June 1995," Massachusetts Institute of Technology, 1995, http:// www.mit.edu/people/mkgray/growth/.

13. John Battelle, *The Search: How Google and Its Rivals Rewrote the Rules of Business and Transformed Our Culture* (New York: Portfolio, 2005).

14. Sergey Brin and Lawrence Page, "Anatomy of a Large-Scale Hypertextual Web Search Engine," Stanford, 1998, http://infolab .stanford.edu/~backrub/google.html.

15. John Battelle, *The Search: How Google and Its Rivals Rewrote the Rules of Business and Transformed Our Culture* (New York: Portfolio, 2005).

16. Ibid.

17. Michael Calore, interview with author, August 8, 2010.

18. This event was attended by the author.

19. Brian Chen, "iPhone Developers Go from Rags to Riches," *Wired*, September 19, 2008, http://www.wired.com/gadgetlab/2008/09/indie-developer/.

20. Brian Chen, "Coder's Half-Million-Dollar Baby Proves iPhone Gold Rush Is Still On," *Wired*, February 12, 2009, http://www.wired.com/gadgetlab/2009/02/shoot-is-iphone/.

21. Tony Dokoupil, "Striking It Rich: Is There an App for That?" *Newsweek*, October 6, 2009, http://www.newsweek.com/2009/10/05/striking-it-rich-is-there-an-app-for-that.html.

22. Dediu, "Apple Has Accepted Nearly 400,000 Apps in 2.5 Years."

23. Vlad Savov, "Apple Ships 120 Million iOS Devices since iPhone's Launch," September 1, 2010, http://www.engadget.com/2010/09/01/Apple-ships-120-million-devices/.

24. Matt Drance, interview with author, August 8, 2010.

25. "1977," Computer History Museum, 2006, http://www.computerhistory.org/timeline/?year=1977.

26. "Apple's App Store Downloads Top 10 Billion," Apple, January 22, 2011, http://www.apple.com/pr/library/2011/01/22appstore.html.

27. John Gruber, "Pound the Quality," DaringFireball, October 27, 2009, http://daringfireball.net/2009/10/pound_the_quality.

28. Steven Levy, *The Perfect Thing: How the iPod Shuffles Commerce, Culture, and Coolness* (New York: Simon & Schuster, 2006).

Chapter 2

1. Phillip Ryu, interview with author, September 3, 2010.

2. Hadley Stern, "Apple Matters Interview: Philip Ryu," Apple-Matters, December 27, 2006, http://www.applematters.com/article/apple-matters-interview-phillip-ryu.

3. David Castelnuovo, interview with author, May 9, 2010.

4. Doug Gross, "New Versions of 'Pocket God' Heading for iPad Versions and Beyond," CNNTech, July 7, 2010, http://articles.cnn

.com/2010-07-07/tech/pocket.god.ipad_1_completely-new-game
-ipad-pc-game?_s=PM:TECH.

5. Rob Murray, interview with author, March 22, 2010.

6. Elinor Mills, "Report: Android App Market Growing Faster than
iPhone Apps," January 20, 2010, http://reviews.cnet.com/8301
-13970_7-20032228-78.html#ixzz1ERLhaXVzr-iphone.

7. Galen Gruman, "Windows Phone 7: Why It's a Disaster for Microsoft," *PC World*, November 23, 2010, http://www.pcworld.idg.com
.au/article/369001/windows_phone_7_why_it_disaster_microsoft.

8. Brian Chen, "A Humbled Microsoft Prepares to Boot Up Windows Phone 7," *Wired*, October 8, 2010, http://www.wired.com/
gadgetlab/2010/10/microsoft-windows-phone7.

9. Michael Gartenberg, interview with author, August 12, 2010.

10. Ross Rubin, interview with author, August 14, 2010.

11. Brian Chen, "Colleges Dream of Paperless, iPad-centric Education," *Wired*, April 5, 2010, http://www.wired.com/gadgetlab/2010/
04/ipad-textbooks.

12. Brian Chen, "Mobile-Inspired Upgrades Define Apple's PC
Strategy," *Wired*, October 20, 2010, http://www.wired.com/gadgetlab/
2010/10/apple-software.

13. Brian Chen, "Mac App Store Provokes Developer Interest,
Concern," *Wired*, October 25, 2010, http://www.wired.com/gadgetlab/
2010/10/mac-app-store.

14. Peter Bright, "Windows 8 Leak: An App Store for Windows,
IE9 Beta in August," ArsTechnica, last updated June 2010, http://
arstechnica.com/microsoft/news/2010/06/leaked-windows-8-slides
-an-app-store-for-windows-ie9-beta-in-august.ars.

Chapter 3

1. Brian Chen, "How the iPhone Could Reboot Education," *Wired*,
December 8, 2009, http://www.wired.com/gadgetlab/2009/12/iphone
-university-abilene.

2. Bill Rankin, interview with author, December 8, 2009.

3. Tyler Sutphen, interview with author, December 8, 2009.

4. Brian Chen, "Microscope Enables Disease Diagnosis with a Cell Phone," *Wired*, May 19, 2008, http://www.wired.com/gadgetlab/tag/malaria-diagnosis-microscope-cellscope-tuberculosis-disease.

5. Wilbur Lam, interview with author, May 19, 2008.

6. Bryan Gardiner, "A Tech Rx for Doctors: The iPhone," *Wired*, March 20, 2008, http://www.wired.com/gadgets/mac/news/2008/03/iphone_doctors.

7. "Funny iPee and Shy Bladder iPhone App Review from KRAPPS," Krapps, Feburary 3, 2009, http://krapps.com/2009/02/03/history-was-made-sort-of.

8. Richard MacManus, "Diabetes Device Connects Wirelessly to iPhone," ReadWriteWeb, March 19, 2009, http://www.readwriteweb.com/archives/diabetes_device_connects_wirelessly_to_iphone.php.

9. Brian Chen, "Digital Contacts Will Keep an Eye on Your Vital Signs," *Wired*, September 10, 2009, http://www.wired.com/gadgetlab/2009/09/ar-contact-lens.

10. Ibid.

11. "Police Investigate Possible Kidnapping," *Boston News*, January 6, 2009, http://www.thebostonchannel.com/news/18421787/detail.html; George Barnes and Danielle Williamson, "Athol Woman and Granddaughter Found in Virginia," *News Telegram*, January 7, 2009, http://www.telegram.com/article/20090107/NEWS/901070289/1116.

12. "Using the GPS for People Tracking," http://www.travelbygps.com/articles/tracking.php.

13. Interview with author, September 13, 2009.

14. Nico Pitney, "Iran Updates (VIDEO): Live-Blogging the Uprising (Saturday, June 13)." *Huffington Post*, June 14, 2009, http://www.huffingtonpost.com/2009/06/14/iran-updates-video-live-b_n_215378.html.

15. Akbar Ganji, "Dear Mr Ban, Heed the Iranian People," *Guardian*, August 12, 2009, http://www.guardian.co.uk/comment isfree/2009/aug/12/iran-ban-ki-moon-protest.

16. John Ham, interview with author, September 10, 2009.

17. Graeme Gerrard, Garry Parkins, Ian Cunningham, Wayne Jones, Samantha Hill, and Sarah Douglas, "National CCTV Strategy," United Kingdom Home Office, October 2007, http://www.statewatch.org/news/2007/nov/uk-national-cctv-strategy.pdf.

Chapter 4

1. "OpenTable, Inc. Announces Third Quarter Financial Results," OpenTable, November 2, 2010, http://press.opentable.com/release detail.cfm?releaseid=526351.

2. "How OpenTable Works for Restaurants," OpenTable (blog), http://blog.opentable.com/2010/how-opentable-works-for-restaurants.

3. Jonathan Wegener, "OpenTable and Restaurant Marketing," *The Back of the Envelope* (blog), February 3, 2009, http://blog.jwegener.com/2009/02/03/opentable-ipo-analysis-restaurant-marketing.

4. Ibid.

5. Mark Pastore, interview with author, January 19, 2011.

6. Suresh Kotha and Debra Glassman, "Starbuck Corporation: Competing in a Global Market," UW Business School, last revised April 3, 2003, http://www.foster.washington.edu/centers/gbc/global businesscasecompetition/Documents/Cases/2003Case.pdf.

7. Jeffrey S. Harrison, Eun-Young Chang, Carina Gauthier, Todd Joerchel, Jorge Nevarez and Meng Wang, "Exporting a North American Concept to Asia: Starbucks in China," May 2005, http://www.entrepreneur.com/tradejournals/article/132354507.html.

8. Kotha and Glassman, "Starbuck Corporation.

9. "Starbucks Posts Its First Quarterly Loss," MSNBC, July 30, 2008, http://www.msnbc.msn.com/id/25936619/ns/business-stocks_and_economy.

10. Laurie J. Flynn, "Starbucks, Awaiting Recovery, Says Profit Fell 77%," *New York Times*, April 29, 2009, http://www.nytimes.com/2009/04/30/business/30sbux.html.

11. "Howard Schultz's Starbucks Memo," *Financial Times*, February 14, 2007, http://www.ft.com/cms/s/0/dc5099ac-c391-11db-9047 -000b5df10621.html#axzz1ERpud8hg.

12. "Starbucks and Chinese Government Announce Yunnan Coffee Industry Investments," Starbucks Newsroom, November 11, 2010, http://news.starbucks.com/article_display.cfm?article_id=464.

13. Scott Fulton, "Apple: Design and Software, Not Hardware, Distinguish Macs from Intel-based PCs," *TG Daily*, January 18, 2006, http://www.tgdaily.com/hardware-features/23832-apple-design-and -software-not-hardware-distinguish-macs-from-intel-based-pcs.

14. "Top Operating System Share Trend," NetMarketShare, accessed February 19, 2011, http://www.netmarketshare.com/os-market -share.aspx?qprid=9&qptimeframe=M&qpsp=131&qpnp=25.

15. Rik Myslewski, "Reliving the Clone Wars," *Macworld*, May 23, 2008, http://www.macworld.com/article/133598/2008/05/macclones .html.

16. "Apple Computer, Inc., Company History," Funding Universe, http://www.fundinguniverse.com/company-histories/Apple -Computer-Inc-Company-History.html.

17. "Paul Thurrott, Apple Computer Halts Cloning, Kills its Biggest Competitor," Windows IT Pro, September 2, 1997, http:// www.windowsitpro.com/article/news2/apple-computer-halts-cloning -kills-its-biggest-competitor.aspx.

18. "Steve Jobs Kills the Clones & Drops the S-Bomb," Zimbio, April 3, 2006, http://www.zimbio.com/Steve+Jobs/articles/nHNi9oktzRI/ Steve+Jobs+Kills+Clones+drops+Bomb.

19. Brian Chen, "Live Blog: Apple Unveils Thinner, Lighter iPad 2," Wired.com, March 2, 2011, http://www.wired.com/gadgetlab/ 2011/03/apple-ipad-liveblog/.

20. "Apple Takes the Lead in the US Smart Phone Market with a 26% Share," Canalys, November 1, 2010, http://www.canalys.com/pr/ 2010/r2010111.html.

21. Brian Chen, "How Microsoft Hit CTRL+ALT+DEL on

Windows Phone," *Wired*, November 8, 2010, http://www.wired.com/gadgetlab/2010/11/making-windows-phone-7/.

22. John Paczkowski, "Microsoft's Steve Ballmer and Ray Ozzie Live at D8," *All Things Digital*, June 3, 2010, http://d8.allthingsd.com/20100603/steve-ballmer-ray-ozzie-session.

23. Brian Chen, "How Microsoft Hit CTRL+ALT+DEL on Windows Phone," *Wired*, November 8, 2010, http://www.wired.com/gadgetlab/2010/11/making-windows-phone-7.

24. Richard N.Langlois and Paul L. Robertson, "Explaining Vertical Integration: Lessons from the American Automobile Industry," *Journal of Economic History* 49, no. 2 (June 1989): 361–75.

25. Ben Worthen, "Companies More Prone to Go 'Vertical,'" *Wall Street Journal*, November 3, 2009, http://online.wsj.com/article/SB125954262100968855.html.

Chapter 5

1. Matthais Gebauer and Frank Patalong, "App 'Censorship' has German Tabloid Fighting Mad," *Spiegel Online*, February 24, 2010, http://www.spiegel.de/international/germany/0,1518,679976,00.html.

2. Brian Chen, "Vanity Fair Columnist Takes His App Rejection Personally," *Wired*, April 28, 2010, http://www.wired.com/epicenter/2010/04/michael-wolff-app.

3. Eric Thomas, "Newber for iPhone Tabled Due to Zero Communication from Apple," Open Letter, March 16, 2009, http://www.mynewber.com/files/EricThomasLetter.pdf.

4. "Steve Jobs' Keynote Address," Apple, June 7, 2010, http://www.apple.com/apple-events/wwdc-2010.

5. Kim Zetter, "TED 2010: Wired for the iPad to Launch by Summer," *Wired*, February 12, 2010, http://www.wired.com/epicenter/2010/02/ted-2010-wired-for-the-ipad-to-launch-by-summer.

6. Nina Link, ed., *The Magazine Handbook 2010/2011* (New York: Magazine Publishers of America, 2010), 21.

7. Peter Kafka, "Hard Labor: Adobe Rebuilds Its Wired Magazine

App to Fit Apple's Flash-Free Agenda," *All Things Digital,* April 30, 2010, http://mediamemo.allthingsd.com/20100430/hard-labor -adobe-rebuilds-its-wired-magazine-app-line-by-line-to-fit-apples -flash-free-agenda/.

8. Steve Jobs, "Thoughts on Flash," Apple, April 2010, http:// www.apple.com/hotnews/thoughts-on-flash.

9. Ryan Singel, "Did FTC Probe Cause Apple to Change App Rules?" *Wired,* September 9, 2010, http://www.wired.com/epicenter/ 2010/09/ftc-apple.

10. "Apple Launches Subscriptions on App Store," Apple, February 15, 2011, http://www.apple.com/pr/library/2011/02/ 15appstore.html.

11. David Carr, "2 Platforms, With 2 Sets of Problems," *New York Times,* February 20, 2011, http://www.nytimes.com/2011/02/21/ business/media/21carr.html.

12. "iOS Developer Program License Agreement," Scribd, http:// www.scribd.com/doc/41213383/iOS-Developer-Program-License -Agreement.

13. Fred von Lohmann, "All Your Apps Are Belong to Apple: The iPhone Developer Program License Agreement," Electronic Frontier Foundation, March 9, 2010, http://www.eff.org/deeplinks/2010/03/ iphone-developer-program-license-agreement-all.

14. Alan Kay, "A Personal Computer for Children of All Ages," Computer History Museum, August 1972, http://history-computer .com/Library/Kay72.pdf.

15. Alan Kay, interview with author, April 16, 2010.

16. John Mcintosh, "Rejecting an App with Foundations in the Dynabook Vision," April 14, 2010, http://mobilewikiserver.com/ Interpreters.html.

17. John Gruber, "A Taste of What's New in the Updated App Store License Agreement and New Review Guidelines," Daring Fire-ball, September 9, 2010, http://daringfireball.net/2010/09/app_store_ guidelines.

18. Brian Chen, "Apple Rejects Kid-Friendly Programming App," *Wired*, April 20, 2010, http://www.wired.com/gadgetlab/2010/04/apple-scratch-app.

19. Mark Pilgrim, "Tinkerer's Sunset," Dive Into Mark, January 29, 2010, http://diveintomark.org/archives/2010/01/29/tinkerers-sunset.

20. Alex Payne, "On the iPad," January 28, 2010, http://al3x.net/2010/01/28/ipad.html.

21. Kai Ryssdal, "'I Partied and I Unlocked the iPhone!'" American Public Media: Market Place, August 24, 2007, http://marketplace.publicradio.org/display/web/2007/08/24/i_partied_and_i_unlocked_the_iphone.

22. James H. Billington, "Statement of the Librarian of Congress Relating to Section 1201 Rulemaking," US Copyright Office, July 26, 2010, http://www.copyright.gov/1201/docs/librarian_statement_01.html.

23. Tim Wu, "The Great American Information Emperors," *Slate*, November 11, 2010, http://www.slate.com/id/2272941.

24. Caroline McCarthy, "Google's Schmidt Resigns from Apple Board," CNET, August 3, 2009, http://news.cnet.com/8301-13579_3-10301612-37.html.

25. John C. Abell, "Google's 'Don't Be Evil' Mantra is 'Bullshit,' Adobe Is Lazy: Apple's Steve Jobs (Update 2)," *Wired*, January 30, 2010, http://www.wired.com/epicenter/2010/01/googles-dont-be-evil-mantra-is-bullshit-adobe-is-lazy-apples-steve-jobs.

26. Priya Ganapati, "Google Introduces Google TV, New Android OS," *Wired*, May 20, 2010, http://www.wired.com/gadgetlab/2010/05/google-introduces-google-tv.

27. Alan Davidson and Tom Tauke, "A Joint Policy Proposal for an Open Internet," Google Public Policy Blog, August 9, 2010, http://googlepublicpolicy.blogspot.com/2010/08/joint-policy-proposal-for-open-internet.html.

28. Karl Bode, "Verizon, Google Announce Their Net Neutrality Solution," DSL Reports, August 9, 2010, http://www.dslreports.com/

shownews/Verizon-Google-Announce-Their-Net-Neutrality-Solution
-109810.

29. "Public Knowledge Says Verizon-Google Agreement Is
'Nothing More Than a Private Agreement between Two Corporate
Behemoths,'" Public Knowledge, August 9, 2010, http://www.public
knowledge.org/public-knowledge-says-verizon-google-agreement
-not.

30. Elia Freedman, "Fighting the Wrong Fight," Elia Insider,
September 14, 2010, http://eliainsider.com/2010/09/14/fighting-the
-wrong-fight.

Chapter 6

1. "Magnus Carlsen: The Official Website," Magnus Carlsen,
http://www.MagnusCarlsen.com.

2. Maik Grossekathöfer, "I Am Chaotic and Lazy," Der Spiegel (English translation), March 15, 2010, http://www.chessbase.com/
newsdetail.asp?newsid=6187.

3. Karl Frisch, "Obama: '24/7 Media . . . Exposes Us to All Kinds
of Arguments, Some of Which Don't Always Rank That High on the
Truth Meter,'" Media Matters, May 9, 2010, http://mediamatters.org/
blog/201005090011.

4. Sarah Lai Stirland, "Propelled by Internet, Barack Obama Wins
Presidency," Wired, November 4, 2008, http://www.wired.com/
threatlevel/2008/11/propelled-by-in.

5. Raven Zachary, "Obama '08 for iPhone," October 2, 2008,
http://raven.me/2008/10/02/obama-08-for-iphone.

6. Brian Chen. "Help! My Smartphone Is Making Me Dumb—
or Maybe Not," Wired, October 4, 2010, http://www.wired.com/
gadgetlab/2010/10/multitasking-studies.

7. Janna Anderson and Lee Rainie, The Future of the Internet IV
(Washington, DC: Pew Internet and American Life Project, February
19, 2010).

8. Eyal Ophir, Clifford Nass, and Anthony D. Wagner, "Cognitive

Control in Media Multitaskers," *Proceedings of the National Academy of Sciences of the United States* 106, no. 37 (September 15, 2009): 15583–87.

9. Mark Liberman, "Are 'Heavy Media Multitaskers' Really Heavy Media Multitaskers?" *Language Log*, September 4, 2010, http://languagelog.ldc.upenn.edu/nll/?p=2607.

10. Nicholas Carr, *The Shallows: What the Internet Is Doing to Our Brains* (New York: W. W. Norton, 2010).

11. Matt Richtel, "Digital Devices Deprive Brain of Needed Downtime," *New York Times*, August 25, 2010, http://www.nytimes.com/2010/08/25/technology/25brain.html.

12. Matt Richtel, "Attached to Technology and Paying a Price," *New York Times*, June 7, 2010, http://www.nytimes.com/2010/06/07/technology/07brain.html.

13. Marjorie Connelly, "More Americans Sense Downside to Being Plugged In," *New York Times*, June 7, 2010, http://www.nytimes.com/2010/06/07/technology/07brainpoll.html.

14. Ophir, Nass, and Wagner, "Cognitive Control in Media Multitaskers."

15. Jason M. Watson and David L. Strayer, "Supertaskers: Profiles in Extraordinary Multitasking Ability," *Psychonomic Bulletin & Review* 17 (2010): 479–85.

16. Zachary, "Obama '08 for iPhone."

17. C. Riels, N. Reed, "Emotion in Motion: Comparing the Effect of a Range of Distractions on Driver Behavior," Transport Research Laboratory, 2011 (forthcoming).

18. Clifford Nass, interview with author, September 25, 2010.

19. Vaughan Bell, "Multi Media, We Don't Need It Do We?" Mind Hacks, August 25, 2009, http://mindhacks.com/2009/08/25/multi-media-we-dont-need-it-do-we.

20. Gary Small and Gigi Vorgan, *iBrain: Surviving the Technological Alteration of the Modern Mind* (New York: William Morrow, 2008).

21. Nicholas Carr, "Author Nicholas Carr: The Web Shatters Fo-

cus, Rewires Brains," *Wired*, May 24, 2010, http://www.wired.com/magazine/2010/05/ff_nicholas_carr.

22. David S. Miall and Teresa Dobson, "Reading Hypertext and the Experience of Literature," *Journal of Digital Information* 2, no. 1 (2001), http://journals.tdl.org/jodi/article/viewArticle/35/37.

23. Carr, "Author Nicholas Carr."

24. Vaughan Bell, "Neuroplasticity Is a Dirty Word," *Mind Hacks*, June 7, 2010, http://mindhacks.com/2010/06/07/neuroplasticity-is-a-dirty-word.

25. Muhammet Demirbilek, "Effects of Interface Windowing Modes on Disorientation in a Hypermedia Learning Environment," *Journal of Educational Multimedia and Hypermedia* 18, no. 4 (2009): 369–83.

26. Erick Schonfeld, "Nobody Predicted the iPad's Growth. Nobody," TechCrunch, January 19, 2011, http://techcrunch.com/2011/01/19/nobody-predicted-ipad-growth.

27. Brian Chen, "Will the iPad Make You Smarter?" *Wired*, July 8, 2010, http://www.wired.com/gadgetlab/2010/07/ipad-interface-studies.

28. Larry Allan Benshoof, Michael Graves, and Simon Hooper, "The Effects of Single and Multiple Window Presentations on Achievement, Instructional Time, Window Use, and Attitudes during Computer-Based Instruction," *Computers in Human Behavior* 11, no. 2 (1995): 261–72.

Chapter 7

1. Choe Sang-Hun, "South Korea Expands Aid for Internet Addiction," *New York Times*, May 28, 2010, http://www.nytimes.com/2010/05/29/world/asia/29game.html; Andrew Salmon, "Couple: Internet Gaming Addiction Led to Baby's Death," CNNWorld, April 1, 2010, http://articles.cnn.com/2010-04-01/world/korea.parents.starved.baby_1_gaming-addiction-internet-gaming-gamingindustry?_s=PM:WORLD.

2. Eric Felten, "Video Game Tort: You Made Me Play You," *Wall*

Street Journal, September 3, 2010, http://online.wsj.com/article/SB10001424052748703369704575461822847587104.html.

3. Robert Weis and Brittany C. Cerankosky, "Effects of Video-Game Ownership on Young Boys' Academic and Behavioral Functioning," *Psychological Science* 21, no. 4 (February 18, 2010): 463–70.

4. R. J. Haier, B. V. Siegel Jr., A. MacLachlan, E. Soderling, S. Lottenberg, and M. S. Buchsbaum, "Regional Glucose Metabolic Changes after Learning a Complex Visuospatial/Motor Task: A Positron Emission Tomographic Study," *Brain Research* 570, nos. 1–2 (January 20, 1992): 134–43.

5. Matthew W. G. Dye, C. Shawn Green, and Daphne Bavelier, "Increasing Speed of Processing with Action Video Games," *Current Directions in Psychological Science* 18, no. 6 (2009): 321–26.

6. S. S. Brady, "Young Adults' Media Use and Attitudes toward Interpersonal and Institutional Forms of Aggression," *Aggressive Behavior* 33, no. 6 (November–December 2007): 519–25.

7. Douglas A. Gentile, Hyekyung Choo, Albert Liau, Timothy Sim, Dongdong Li, Daniel Fung, and Angeline Khoo, "Pathological Video Game Use Among Youths: A Two-Year Longitudinal Study," *Pediatrics* 127, no. 3 (February, 1, 2011): 319–29.

8. Daphne Bavelier, C. Shawn Green, Matthew W. G. Dye, "Children, Wired: For Better and For Worse," *Neuron* 67, no. 5 (September 9, 2010): 692–701.

9. Neils Clark and P. Shauvan Scott, *Game Addiction: The Experience and the Effects* (Jefferson, NC: McFarland, 2009).

10. Jason Morris, Jeff Reese, Richard Beck, Charles Mattis, "Facebook as a Predictor of Retention at a Private 4-Year Institution," *Journal of College Student Retention* 11, no. 3 (2010): 311–22.

11. Sarah H. Konrath, Edward H. O'Brien, Courtney Hsing, "Changes in Dispositional Empathy in American College Students Over Time: A Meta-Analysis," *Personality and Social Psychological Review* (online August 5, 2010).

12. Chris Matyszczyk, "'Character Amnesia' Hitting Gear-

Obsessed Kids," CNET News, August 26, 2010, http://news.cnet.com/8301-17852_3-20014834-71.html.

13. D. Lin, C. McBride-Chang, H. Shu, Y. P. Zhang, H. Li, J. Zhang, D. Aram, and I. Levin, "Small Wins Big: Analytic Pinyin Skills Promote Chinese Word Reading," *Psychological Science* 21, no. 8 (August 2010): 1117–22.

14. Amanda Lenhart, Sousan Arafeh, Aaron Smith, and Alexandra Macgill, *Writing, Technology, and Teens* (Washington DC: Pew Internet and American Life Project, 2008).

15. US Department of Education, Institute of Education Sciences, National Center for Education Statistics, National Assessment of Educational Progress (NAEP), "1998, 2002, and 2007 Writing Assessments," Nation's Report Card, http://nationsreportcard.gov/writing_2007/w0003.asp.

16. "Table 425: Public Schools and Instructional Rooms with Internet Access, by Selected School Characteristics: Selected Years, 1994 through 2005," in *Digest of Educational Statistics: 2009*, National Center for Education Statistics, April 2010, http://nces.ed.gov/programs/digest/d09/tables/dt09_425.asp?referrer=list.

17. James Paul Gee, *What Video Games Have to Teach Us About Learning and Literacy* (New York: Palgrave Macmillan, 2003), 48.

18. Ibid.

19. Timothy Ferriss, *The 4-Hour Work Week: Escape the 9–5, Live Anywhere and Join the New Rich*, Kindle ed. (New York: Crown Archetype, 2009).

20. Ibid., 1394.

21. Gloria Mark, Victor M. Gonzalez, and Justin Harris, "No Task Left Behind? Examining the Nature of Fragmented Work," *Computer Human Interaction Conference*, April 2–7, Portland, OR.

22. Jonathan B. Spira and Joshua B. Feintuch, *The Cost of Not Paying Attention: How Interruptions Impact Knowledge Workers' Productivity* (New York: Basex, 2005).

23. Mary Madden and Sydney Jones, *Networked Workers* (Washing-

ton, DC: Pew Internet and American Life Project, September 24, 2008), http://www.pewinternet.org/Reports/2008/Networked -Workers/1-Summary-of-Findings.aspx.

24. "Freedom to Surf: Workers More Productive If Allowed to Use the Internet for Leisure," The University of Melbourne, April 2, 2009, http://uninews.unimelb.edu.au/news/5750.

25. Vaughan Bell, "Don't Touch That Dial!" Slate, February 15, 2010, http://www.slate.com/id/2244198.

Chapter 8

1. "24 Hours: Unplugged," International Center for Media and the Public Agenda, http://withoutmedia.wordpress.com.

2. Paul Atchley, Stephanie Atwood, and Aaron Boulton, "The Choice to Text and Drive in Younger Drivers: Behavior May Shape Attitude," *Accident Analysis and Prevention* 43, no. 1 (2010): 134–42.

3. Anita Smith and Kipling D. Williams, "R U There? Ostracism by Cell Phone Text Messages," *Group Dynamics* 8, no. 4 (2004): 291– 301.

4. Kevin Kelly, "The Technium: Amish Hackers," KK, February 10, 2009, http://www.kk.org/thetechnium/archives/2009/02/amish_ hackers_a.php.

5. Ibid.

Chapter 9

1. Echo Metrix, Inc., "PULSE Is Tapped in to What Drives the $190B Teen Market," Marketwire, June 29, 2009, http://www. marketwire.com/press-release/PULSE-Is-Tapped-Into-What-Drives -the-190B-Teen-Market-1010646.htm.

2. Ibid.

3. Kimberly Nguyen and Marc Rotenberg, "Before the Federal Trade Commission in the Matter of Echometrix, Inc.: Complaint, Request for Investigation, Injunction, and Other Relief," Electronic Privacy Information Center, September 25, 2009.

4. "Echometrix to Pay $100,000 Fine and Stop Unfair Practices in Settlement with NY AG," Epic.org, September 17, 2010, http://epic .org/2010/09/echometrix-to-pay-100000-fine.html.

5. Simson Garfinkle, "Software That Can Spy on You," Salon, June 15, 2000, http://www.salon.com/technology/col/garf/2000/06/15/ brodcast.

6. *Robbins et al. vs. Lower Merion School District et al.*, US District Court for the Eastern District of Pennsylvania, February 11, 2010.

7. Associated Press, "FBI Probing School Webcam Spy Case," republished at CBS News, February 19, 2010, http://www.cbsnews .com/stories/2010/02/19/tech/main6223192.shtml.

8. Larry Magid, "Webcams Spying: It's More Than Software," February 23, 2010, http://www.cbsnews.com/stories/2010/02/23/ eveningnews/techtalk/main6233519.shtml.

9. Electronic Privacy Information Center, "2010 Children's Online Privacy Protection Act Rule Review," FTC Matter No. P104503, July 9, 2010, http://www.ftc.gov/os/comments/copparulerev2010/ 547597-00061-54978.pdf.

10. Adrian Chen, "GCreep: Google Engineer Stalked Teens, Spied on Chats (Updated)," *Gawker*, September 14, 2010, http://gawker .com/#!5637234.

11. Peter Kafka, "Apple CEO at D8: The Full, Uncut Interview (with Walt Mossberg and Kara Swisher)," *All Things Digital*, June 7, 2010.

12. "TRUSTe Frequently Asked Questions," http://www.truste .com/about_TRUSTe/faqs.html#3rdparty.

13. David Barnard, "Anti-Competitive AND Potentially Creepy," @drbarnard, June 2010, http://davidbarnard.com/post/684540619/ anti-competitive-and-potentially-creepy.

14. Marc Rotenberg, John Verdi, Ginger McCall, and Veronica Louie, "Before the Federal Trade Commission in the Matter of Facebook, Inc.: Complaint, Request for Investigation, Injunction, and Other Relief," Electronic Privacy Information Center, May 5, 2010, http://epic.org/privacy/facebook/EPIC_FTC_FB_Complaint.pdf.

15. Guilbert Gates, "Facebook Privacy: A Bewildering Tangle of Options," *New York Times*, May 12, 2010, http://www.nytimes.com/interactive/2010/05/12/business/face; "COMPLAINT: In the Matter of Facebook," Epic.org, May 5, 2010, http://epic.org/privacy/facebook/EPIC_FTC_FB_Complaint.pdf book-privacy.html.

16. Mark Zuckerberg, "From Facebook, Answering Privacy Concerns with New Settings," *The Washington Post*, May 24, 2010, http://www.washingtonpost.com/wp-dyn/content/article/2010/05/23/AR2010052303828.html.

17. "Does Google Care about Privacy and Trust?" TRUSTe (blog), May 30, 2008, http://www.truste.com/blog/?p=85.

18. Brian Chen, "iPhone Can Take Screenshots of Anything You Do," *Wired*, September 11, 2008, http://www.wired.com/gadgetlab/2008/09/hacker-says-sec.

19. Jim Harper, "It's Modern Trade: Web Users Get as Much as They Give," *The Wall Street Journal*, August 7, 2010, http://online.wsj.com/article/SB10001424052748703748904575411530096840958.html.

20. Will Moffett, interview with author, May 27, 2010.

Chapter 10

1. Brian Chen, "Blind Photographers Use Gadgets to Realize Artistic Vision," *Wired*, July 16, 2009, http://www.wired.com/gadgetlab/2009/07/blind-photographers.

2. Brian Chen, "If You're Not Seeing Data, You're Not Seeing," *Wired*, August 25, 2009, http://www.wired.com/gadgetlab/2009/08/augmented-reality.

3. David Mizell, "Boeing's Wire Bundle Assembly Project," in *Fundamentals of Wearable Computers and Augmented Reality*, ed. Woodrow Barfield and Thomas Caudell (Mahwah, NJ: Lawrence Erlbaum Associates, 2001).

4. Ronald Azuma, "A Survey of Augmented Reality," Hughes Research Laboratories, August 1997, http://www.cs.unc.edu/~azuma/ARpresence.pdf.

5. Raimo van der Klein, interview with author, August 20, 2009.

6. Brian Chen, "Augmented Reality App Identifies Strangers with Camera," *Wired*, February 24, 2010, http://www.wired.com/gadgetlab /2010/02/augmented-reality-app-identifies-strangers-with-camera -software.

7. "ARhrrr!" Augmented Environments Lab, http://www.augmented environments.org/lab/research/handheld-ar/arhrrr.

8. Kai Ryssdal, "District 9 Creates Augmented Reality." American Public Media's Marketplace, August 14, 2009, http://marketplace. publicradio.org/display/web/2009/08/14/pm-district-9-q.

9. Brian Selzer, interview with author, August 20, 2009.

10. Hrvoje Benko, "Collaborative Visualization of an Archeological Excavation," Columbia University, http://graphics.cs.columbia.edu/ projects/ArcheoVis.

11. Jackie Fenn and David McCoy, "Wearable Computing Pioneer Thad Starner," *Gartner*, http://www.gartner.com/research/fellows/ asset_196289_1176.jsp.

12. "The 44th President: Inauguration," CNNPolitics, http://www. cnn.com/SPECIALS/2009/44.president/inauguration/themoment/.

acknowledgments

I've argued that we're living in a very special time where data and technology can help us in so many ways — but I couldn't have finished this project without the support of many friends and colleagues.

First and foremost, thank you to Megan Geuss, who did research and fact-checking for Always On: without you, I would never have made it to the finish line.

David Fugate, my agent at LaunchBooks — an epic thank you for helping me pound this complex topic into a solid book proposal, and for going above and beyond your job description looking out for me.

Huge thanks to my coworkers at Wired.com, for their support and patience throughout my "Year of the Book": Dylan Tweney, Betsy Mason, Amy Ashcroft, David Kravets, Jon Snyder, Evan Hansen, Michael Calore, Ryan Singel, and Erik Malinowski.

A special thank you to author Leander Kahney, who spun around in his chair one day in the Wired newsroom and planted the idea of writing a book in my head. And big thanks to Rana Sobhany, Phillip Ryu and Alexis Madrigal for motivating me and offering me guidance throughout this challenging project.

Much love to my friends who were there for me from start to finish: Pamela Wong, David Lee, Peter Hamilton, Peter Nguyen,

Tracy Young, Debra Kahn, Heather Kelly, Rachael Bogert, Rosa Golijan, Rose Roark, Andreas Schobel, Alex Krawiec, Amy Zimmerman, Jenn de la Vega, Stephanie Hammon, and Deborah Feuer.

And with that said, I'm signing off.

index